Climate
AND SOCIAL STRESS

Implications for Security Analysis

Committee on Assessing the Impacts of Climate Change on
Social and Political Stresses

John D. Steinbruner, Paul C. Stern, and Jo L. Husbands, *Editors*

Board on Environmental Change and Society

Division of Behavioral and Social Sciences and Education

NATIONAL RESEARCH COUNCIL
OF THE NATIONAL ACADEMIES

THE NATIONAL ACADEMIES PRESS
Washington, D.C.
www.nap.edu

THE NATIONAL ACADEMIES PRESS   500 Fifth Street, NW   Washington, DC 20001

NOTICE: The project that is the subject of this report was approved by the Governing Board of the National Research Council, whose members are drawn from the councils of the National Academy of Sciences, the National Academy of Engineering, and the Institute of Medicine. The members of the committee responsible for the report were chosen for their special competences and with regard for appropriate balance.

This study was supported by the U.S. intelligence community. Any opinions, findings, conclusions, or recommendations expressed in this publication are those of the author(s) and do not necessarily reflect the views of the organizations or agencies that provided support for the project.

International Standard Book Number-13: 978-0-309-27856-0
International Standard Book Number-10: 0-309-27856-2

Library of Congress Cataloging-in-Publication data are available from the Library of Congress.

Additional copies of this report are available for sale from the National Academies Press, 500 Fifth Street, NW, Keck 360, Washington, DC 20001; (800) 624-6242 or (202) 334-3313; http://www.nap.edu.

Copyright 2013 by the National Academy of Sciences. All rights reserved.

Printed in the United States of America

Suggested Citation: National Research Council. (2013). *Climate and Social Stress: Implications for Security Analysis*. Committee on Assessing the Impacts of Climate Change on Social and Political Stresses, J.D. Steinbruner, P.C. Stern, and J.L. Husbands, Eds. Board on Environmental Change and Society, Division of Behavioral and Social Sciences and Education. Washington, DC: The National Academies Press.

THE NATIONAL ACADEMIES
Advisers to the Nation on Science, Engineering, and Medicine

The **National Academy of Sciences** is a private, nonprofit, self-perpetuating society of distinguished scholars engaged in scientific and engineering research, dedicated to the furtherance of science and technology and to their use for the general welfare. Upon the authority of the charter granted to it by the Congress in 1863, the Academy has a mandate that requires it to advise the federal government on scientific and technical matters. Dr. Ralph J. Cicerone is president of the National Academy of Sciences.

The **National Academy of Engineering** was established in 1964, under the charter of the National Academy of Sciences, as a parallel organization of outstanding engineers. It is autonomous in its administration and in the selection of its members, sharing with the National Academy of Sciences the responsibility for advising the federal government. The National Academy of Engineering also sponsors engineering programs aimed at meeting national needs, encourages education and research, and recognizes the superior achievements of engineers. Dr. Charles M. Vest is president of the National Academy of Engineering.

The **Institute of Medicine** was established in 1970 by the National Academy of Sciences to secure the services of eminent members of appropriate professions in the examination of policy matters pertaining to the health of the public. The Institute acts under the responsibility given to the National Academy of Sciences by its congressional charter to be an adviser to the federal government and, upon its own initiative, to identify issues of medical care, research, and education. Dr. Harvey V. Fineberg is president of the Institute of Medicine.

The **National Research Council** was organized by the National Academy of Sciences in 1916 to associate the broad community of science and technology with the Academy's purposes of furthering knowledge and advising the federal government. Functioning in accordance with general policies determined by the Academy, the Council has become the principal operating agency of both the National Academy of Sciences and the National Academy of Engineering in providing services to the government, the public, and the scientific and engineering communities. The Council is administered jointly by both Academies and the Institute of Medicine. Dr. Ralph J. Cicerone and Dr. Charles M. Vest are chair and vice chair, respectively, of the National Research Council.

www.national-academies.org

COMMITTEE ON ASSESSING THE IMPACTS OF CLIMATE CHANGE ON SOCIAL AND POLITICAL STRESSES

JOHN D. STEINBRUNER (*Chair*), Professor of Public Policy, University of Maryland; Director, Center for International and Security Studies at Maryland
OTIS B. BROWN, Director, Cooperative Institute for Climate and Satellites, North Carolina State University
ANTONIO J. BUSALACCHI, JR., Director, Earth System Science Interdisciplinary Center, University of Maryland; Professor, Department of Atmospheric and Oceanic Science
DAVID EASTERLING, Chief, Scientific Services Division, National Climatic Data Center, National Oceanic and Atmospheric Administration, Asheville, NC
KRISTIE L. EBI, Consulting Professor, Department of Medicine, Stanford University
THOMAS FINGAR, Oksenberg–Rohlen Distinguished Fellow and Senior Scholar, Freeman Spogli Institute for International Studies, Stanford University
LEON FUERTH, Distinguished Research Fellow, National Defense University; Research Professor of International Affairs, George Washington University; Founder and Director, Project on Forward Engagement
SHERRI GOODMAN, Senior Vice President, General Counsel, and Corporate Secretary, CNA Analysis and Solutions, Alexandria, VA; Executive Director, CNA Military Advisory Board
ROBIN LEICHENKO, Associate Professor, Department of Geography, Rutgers University
ROBERT J. LEMPERT, Director, Frederick S. Pardee Center for Longer Range Global Policy and the Future Human Condition, RAND Corporation, Santa Monica, CA
MARC LEVY, Deputy Director, Center for International Earth Science Information Network, Earth Institute, Columbia University
DAVID LOBELL, Assistant Professor, Environmental Earth System Science, Stanford University; Center Fellow, Program on Food Security and the Environment, Stanford University
RICHARD STUART OLSON, Director of Extreme Event Research and Professor, Department of Politics and International Relations, Florida International University
RICHARD L. SMITH, Director, Statistical and Applied Mathematical Sciences Institute, Research Triangle Park, NC

PAUL C. STERN, *Study Director*
JO L. HUSBANDS, *Scholar*
ALICIA JARAMILLO-UNDERWOOD, *Senior Program Assistant*
MARY ANN KASPER, *Senior Program Assistant*

BOARD ON ENVIRONMENTAL CHANGE AND SOCIETY

RICHARD H. MOSS (*Chair*), Senior Staff Scientist, Joint Global Change Research Institute, College Park, MD
ARUN AGRAWAL, Research Associate Dean, School of Natural Resources and Environment, University of Michigan, Ann Arbor
JOSEPH ARVAI, Svare Chair in Applied Decision Research, University of Calgary
ANTHONY BEBBINGTON, Higgins Professor of Environment and Society, Director of the Graduate School of Geography, Clark University
WILLIAM CHANDLER, President, Transition Energy, Annapolis, MD
F. STUART CHAPIN, III, Professor, University of Alaska–Fairbanks
RUTH DEFRIES, Denning Professor of Sustainable Development, Department of Ecology, Evolution, and Environmental Biology, Columbia University
KRISTIE L. EBI, Consulting Professor, Department of Medicine, Stanford University
MARIA CARMEN LEMOS, Associate Professor, School of Natural Resources and Environment, University of Michigan, Ann Arbor
DENNIS OJIMA, Senior Research Scientist, Natural Resource Ecology Laboratory, Colorado State University
JONATHAN OVERPECK, Co-Director, Institute of the Environment, University of Arizona
STEPHEN POLASKY, Professor of Ecological/Environmental Economics, Department of Applied Economics, University of Minnesota
J. TIMMONS ROBERTS, Director, Center for Environmental Studies, Professor of Sociology and Environmental Studies, Center for Environmental Studies, Brown University
JAMES L. SWEENEY, Professor of Management Science and Engineering, Stanford University
GARY W. YOHE, Woodhouse/Sysco Professor of Economics, Department of Economics, Wesleyan University

MEREDITH A. LANE, *Board Director*
PAUL C. STERN, *Senior Scholar*
MARY ANN KASPER, *Senior Program Assistant*

Preface

Core features of the climate change situation are known with confidence. The greenhouse effect associated with the carbon dioxide molecule has been measured, as has the dwell time of that molecule and its concentration in the atmosphere. We also know that the rate at which carbon dioxide is currently being added to the atmosphere substantially exceeds the natural rate that prevailed before the rise of human societies. That means that a large and unprecedentedly rapid thermal impulse is being imparted to the earth's ecology that will have to be balanced in some fashion. We know beyond reasonable doubt that the consequences will be extensive. We do not, however, know the timing, magnitude, or character of those consequences with sufficient precision to make predictions that meet scientific standards of confidence.

In principle the thermal impulse could be mitigated to a degree that would presumably preserve the current operating conditions of human societies, but the global effort required to do that is not being undertaken and cannot be presumed. As a practical matter, that means that significant burdens of adaptation will be imposed on all societies and that unusually severe climate perturbations will be encountered in some parts of the world over the next decade with increasing frequency and severity thereafter. There is a compelling reason to presume that specific failures of adaptation will occur with consequences more severe than any yet experienced, severe enough to compel more extensive international engagement than has yet been anticipated or organized.

This report has been prepared at the request of the U.S. intelligence community with these circumstances in mind. It summarizes what is currently known about the security effects of climate perturbations, admitting the inherent complexities and the very considerable uncertainties involved. But under the presumption that these effects will be of increasing significance, it outlines the monitoring activities that the intelligence community should be developing in support of improved anticipation, more effective prevention efforts, and more decisive emergency reaction when that becomes necessary.

The report was prepared by the members of the committee, all of whom helped shape the assessment presented and many of whom drafted elements of the text. The burden of constructing a coherent whole from individual contributions fell primarily to Paul Stern and Jo Husbands as the principal editors of the report. Alicia Jaramillo-Underwood and Mary Ann Kasper provided essential administrative support. National Research Council Fellow Andrei Israel and intern Zafar Imran provided research support and assisted in the preparation of parts of the text. I am personally grateful for all of these contributions.

John D. Steinbruner, *Chair*
Committee on Assessing the Impacts of
Climate Change on Social and Political Stresses

Acknowledgments

This report has been reviewed in draft form by individuals chosen for their diverse perspectives and technical expertise, in accordance with procedures approved by the National Academies' Report Review Committee. The purpose of this independent review is to provide candid and critical comments that will assist the institution in making its published report as sound as possible and to ensure that the report meets institutional standards for objectivity, evidence, and responsiveness to the study charge. The review comments and draft manuscript remain confidential to protect the integrity of the process.

We wish to thank the following individuals for their review of this report: Marc F. Bellemare, Sanford School of Public Policy, Duke University; Andrew Brown, Jr., Innovation and Technology Office, Delphi Corporation, Troy, Michigan; Jared L. Cohon, Office of the President, Carnegie Mellon University; Geoff Dabelko, Environmental Change and Security Program, Woodrow Wilson Center; Delores M. Etter, Caruth Institute for Engineering Education, Southern Methodist University; John Gannon, BAE Systems, Arlington, Virginia; James R. Johnson, (retired) Minnesota Mining and Manufacturing Company, Oak Park Heights, Minnesota; John E. Kutzbach, Center for Climatic Research, University of Wisconsin–Madison; Monty G. Marshall, Center for Global Policy, George Mason University and Center for Systemic Peace, Societal-Systems Research, Inc.; Dennis Ojima, Ecosystem Science and Sustainability, Warner College of Natural Resources, Colorado State University; Reto Ruedy, NASA Goddard Institute for Space Studies; and Philip A. Schrodt, Department of Political Science, Pennsylvania State University.

Although the reviewers listed above have provided many constructive comments and suggestions, they were not asked to endorse the conclusions or recommendations, nor did they see the final draft of the report before its release. The review of this report was overseen by Warren M. Washington, Climate Change Research Section, Climate and Global Dynamics Division, National Center for Atmospheric Research, and Thomas J. Wilbanks, Environmental Sciences Division, Oak Ridge National Laboratory. Appointed by the National Academies, they were responsible for making certain that an independent examination of this report was carried out in accordance with institutional procedures and that all review comments were carefully considered. Responsibility for the final content of this report rests entirely with the authoring committee and the institution.

Contents

Summary   1

1   **Climate Change as a National Security Concern**   15
Potential Climate–Security Connections, 17
Increasing Risks of Disruptive Climate Events, 21
The Focus of This Study, 30
Structure of the Report, 33

2   **Climate Change, Vulnerability, and National Security:
A Conceptual Framework**   35
Connections Between Climate Events and National Security, 36
Implications of the Conceptual Framework, 43
Strategies for Security Analysis, 48

3   **Potentially Disruptive Climate Events**   53
The Science of Climate Projection, 54
Abrupt Climate Change, 58
Single Extreme Events, 61
Clusters of Extreme Events, 68
Sequences of Events, 70
Global System Shocks, 71
Surprises Arising from Poorly Resolved Climate Dynamics, 72
Conclusions and Recommendations, 73

4	How Climate Events Can Lead to Social and Political Stresses	75

Local and Distant Effects, 76
Exposures, 82
Susceptibility to Harm from Climate Events, 84
Coping, Response, and Recovery, 87
Conclusions and Recommendations, 91

5	Climate Events and National Security Outcomes	97

Water, Food, and Health Security, 98
Humanitarian Crises, 111
Disruptive Migration, 112
Severe Political Instability and State Failure, 117
Interstate and Intrastate Conflict and Violence, 125
Conclusions and Recommendations, 134

6	Methods for Assessing National Security Threats	139

What Should Be Monitored and Why, 140
Challenges of Monitoring, 143
A Strategy for Monitoring, 153
An Approach to Anticipating Risks, 158

References 161

Appendixes

A	Committee Member and Staff Biographies	179
B	Briefings Received by the Committee	187
C	Method for Developing Figure 3-1	189
D	Statistical Methods for Assessing Probabilities of Extreme Events	193
E	Foundations for Monitoring Climate–Security Connections	203

Summary

The U.S. intelligence community is expected to provide indicators and warnings of a wide variety of security threats—not only risks of international wars that might threaten U.S. interests or require a U.S. military response, but also risks of violent subnational conflicts in countries of security concern, risks to the stability of states and regions, and risks of major humanitarian disasters in key regions of the world. This intelligence mission requires the consideration of activities and processes anywhere in the world that might lead, directly or indirectly, to significant risks to U.S. national security.

In recent years, with the accumulation of scientific evidence indicating that the global climate is moving outside the bounds of past experience and can be expected to put new stresses on societies around the world, the U.S. intelligence and security communities have begun to examine a variety of plausible scenarios through which climate change might pose or alter security risks. In 2010, as part of its ongoing work with the National Academy of Sciences/National Research Council (NAS/NRC) on issues related to climate and security, the U.S. intelligence community asked the NAS/NRC to organize the study whose results are described in this report.

The central purpose of the study, as defined in its statement of task, was "to evaluate the evidence on possible connections between climate change and U.S. national security concerns and to identify ways to increase the ability of the intelligence community to take climate change into account in assessing political and social stresses with implications for U.S. national security." The study committee was tasked to "focus on several broad questions, such as: What are the major social and political factors affecting the relationship between climate change and outcomes relevant to U.S. national security? What is the basis for this knowledge and how strong is it? What research and measurement strategies would strengthen the basis for this knowledge?" In response to this charge, this report presents a conceptual framework for addressing such issues, offers an evaluation of the available evidence, identifies key factors linking climate change phenomena to security concerns, and offers conclusions and recommendations related to: (a) improving understanding of climate–security linkages; (b) improving monitoring and analysis of the factors linking climate change to social and political stresses and to security risks; and (c) improving the ability to anticipate potential security risks arising from climate phenomena.

As the study developed, and upon consultation with the study's sponsors, we focused our efforts in three specific ways. First, we focused on social and political stresses outside the United States because such stresses are the main focus of the intelligence community. Second, we concentrated on security risks that might arise from situations in which climate events (e.g., droughts, heat waves, or storms) have consequences that exceed the capacity of affected countries or populations to cope and respond. This focus led us to exclude, for example, climate events that might directly affect the ability of the U.S. military to conduct its missions or that might contribute directly to international competition or conflict (e.g., over sea lanes or natural resources in the Arctic). We also excluded the security implications of policies that countries might undertake to protect themselves from perceived threats of climate change (e.g., geoengineering to reduce global warming or buying foreign agricultural land to ensure domestic food supplies). These kinds of climate–security connections could prove highly significant and deserve further study and analysis. They could also interact with the connections that are our main focus; for example, an action such as buying foreign agricultural land might go almost unnoticed at first, only creating a crisis when the country where the land is located experiences a crop failure it cannot manage with imports. Third, we concentrated on the relatively near term by emphasizing climate-driven security risks that call for action by the intelligence community within the coming decade either to respond to security threats or to anticipate them.

Although these choices of focus helped bound our study, they left it with some notable limitations. Climate change is a global and a long-term phenomenon. Events within the United States and those outside the country affect each other, indirect links between climate and conflict can be related to direct ones, and the effects of climate change will not stop beyond a 10-year horizon and, in fact, can be expected to increase at an increasing rate. Thus a complete security analysis should project the risks of climate change beyond the next decade in order to inform U.S. government security policy choices in the near term that will prepare the nation for events in later decades.

Our study includes the full range of potentially disruptive events that are becoming more likely because of climate change, whether or not a particular event can be unequivocally attributed to human-caused climate change rather than to natural variation. We made this choice because any such climate events can become disruptive and create a need for U.S. government action regardless of whether they can at this time be uniquely attributed to anthropogenic climate change.

KNOWLEDGE ABOUT CLIMATE–SECURITY CONNECTIONS

Anthropogenic climate change can reasonably be expected to increase the frequency and intensity of a variety of potentially disruptive environmental events—slowly at first, but then more quickly. Some of this change is already discernible. Many of these events will stress communities, societies, governments, and the globally integrated systems that support human well-being. Science is unlikely ever to be able to predict the timing, magnitude, and precise location of these events a decade in advance, but much is already known that can inform security analysis, including details about the character of events that are becoming more likely and about the general trajectory of increasing risk.

Conclusion 3.1[1]: *Given the available scientific knowledge of the climate system, it is prudent for security analysts to expect climate surprises in the coming decade, including unexpected and potentially disruptive single events as well as conjunctions of events occurring simultaneously or in sequence, and for them to become progressively more serious and more frequent thereafter, most likely at an accelerating rate. The climate surprises may affect particular regions or globally integrated systems, such as grain markets, that provide for human well-being.*

The conjunctions of events will likely include clusters of apparently unrelated climate events occurring closely in time, although perhaps widely separated geographically, which actually do have common causes; sequences or cascades of events in which a climate event precipitates a series of other physical or biological consequences in unexpected ways; and disruptions of globally connected systems, such as food markets, supply chains for strategic commodities, or global public health systems. The surprises are likely to appear first as unusually severe extensions of familiar experience.

---

[1]Conclusions and recommendations are numbered to indicate the chapter where they appear and their ordering within that chapter.

Some of them are likely to be felt in regions remote from where the actual climate events take place. It is prudent to expect that some of these events will create or exacerbate conditions affecting U.S. national security.

It makes sense for the intelligence community to apply a scenario approach in thinking about potentially disruptive events that are expectable but not truly predictable. For example, when climate models disagree about the direction of a climate trend even when the fundamental science strongly suggests that change is likely, it may make sense to consider the security implications of two or more plausible trends as a way to anticipate risks.

> **Conclusion 4.1:** *The overall risk of disruption to a society from a climate event is determined by the interplay among several factors: event severity, exposure of people or valued things, and the vulnerability of those people or things, including susceptibility to harm and the effectiveness of coping, response, and recovery. Exposure and vulnerability may pertain to the direct effects of a climate event or to effects mediated by globalized systems that support the well-being of the society.*

The security risks are unlikely to be anticipated by looking only at climate trends and projections. Each of the factors affecting disruption is changing, and several are changing in ways that can be projected with some confidence for a decade or more at the country level or below. Because risk reflects the interactions among these factors and not only the magnitude of climate events, events of a magnitude that has not been disruptive in the past can cause major social and political disruption if exposure and susceptibility are sufficiently great and response is inadequate or widely seen as such. The other side of this coin is that unprecedentedly large climate events do not necessarily lead to security threats if actions have been taken to reduce exposure or susceptibility or increase coping capacity and if authorities are seen to be actively responding to events.

> **Conclusion 4.2:** *To understand how climate change may create social and political stresses with implications for U.S. national security, it is essential for the intelligence community to understand adaptation and changes in vulnerability to climate events and their consequences in places and systems of concern, including susceptibility to harm and the potential for effective coping, response, and recovery. This understanding must be integrated with understanding of changes in the likelihoods of occurrence of climate events.*

Knowledge from several scientific fields provides useful general insights about the components of vulnerability and how they shape the effects of climate events on social and political systems. Much remains to be done, however, to advance this knowledge and make it operational for assessing the risks of climate change to social and political systems in particular places.

> **Conclusion 5.1:** *It is prudent to expect that over the course of a decade some climate events—including single events, conjunctions of events occurring simultaneously or in sequence in particular locations, and events affecting globally integrated systems that provide for human well-being—will produce consequences that exceed the capacity of the affected societies or global systems to manage and that have global security implications serious enough to compel international response. It is also prudent to expect that such consequences will become more common further in the future.*

> **Conclusion 5.2:** *The links between climate events and security outcomes are complex, contingent, and not understood nearly well enough to allow for prediction. However, the key linkages, as with societal disruptions, seem prominently to involve (a) exposures to potentially disruptive events directly or through globally integrated systems affecting human well-being and (b) vulnerabilities (i.e., susceptibility to harm and the effectiveness of coping, response, and recovery efforts). In addition, security outcomes depend on the reactions of social and political systems to actual or perceived inadequacies of response.*

Available knowledge of climate–security connections that feature societal vulnerabilities indicates that security analysis needs to develop more nuanced understanding of the conditions—largely, social, political, and economic conditions—under which particular climate events are and are not likely to lead to particular kinds of social and political stresses and under which such events and responses to them are and are not likely to lead to significant security threats.

The empirical knowledge base on the connections between extreme events and political instability or violence also suggests some hypotheses that are worthy of further examination. For example, available knowledge is consistent with a model in which the link of climate events to the potential for significant violence, conflict, or breakdown depends on these factors:

- the nature, breadth or concentration, and depth of pre-existing social and political grievances and stresses;
- the nature, breadth or concentration, and depth of the immediate impacts of the climate event;

- the socioeconomic, geographic, racial, ethnic, and religious profiles of the most exposed groups or subpopulations, as well as their susceptibilities and coping capacities;
- the ability and willingness of the incumbent government and its internal and external supporters to devise, publicize, and implement effective, transparent, and equitable short-term emergency response and then longer-term recovery plans;
- the extent to which emergent or established anti-government or anti-regime movements or groups are able to take strategic or tactical advantage of grievances or problems related to responses to the event;
- the type, breadth, and depth of legitimacy and support for authorities, the government, the regime, or the nation–state; and
- the coercive and repressive capacities of the government and its willingness and ability to engage and carry out repression.

TOWARD IMPROVED MONITORING, ANALYSIS, AND ANTICIPATION

The intelligence and national security communities are not the only parts of the U.S. government that need improved understanding of vulnerabilities to climate change to achieve their goals, and the U.S. government is not the only actor that has this need. Such improved understanding is among the objectives of the many federal scientific agencies concerned with climate change and will be valuable to the various federal, state, local, private-sector, and international organizations concerned with improving adaptation to climate change, reducing potential damage from climate events, and exploiting potential opportunities related to climate change. These shared needs for knowledge suggest that knowledge development is best pursued as a cooperative activity involving many organizations.

A recent report of the Defense Science Board (Defense Science Board, 2011) emphasized the need for federal interagency cooperation in dealing with issues of adaptation to climate change. It called for "a structure and process for coordination to more effectively leverage the efforts to address global problems" and "a whole of government approach on regional climate change adaptation with a focus on promoting climate change resilience and maintaining regional stability." We agree with the need for a whole-of-government approach and note that the effort should include improved knowledge and monitoring of changing vulnerabilities as well as of climate trends.

Within the U.S. government, the entity charged with developing fundamental knowledge about climate vulnerabilities is the U.S. Global Change Research Program (USGCRP). One of the five scientific objectives in its strategic plan for 2012–2021 is to "[a]dvance understanding of the vulnerability and resilience of integrated human–natural systems and enhance the usability of scientific knowledge in supporting responses to global change" (U.S. Global Change Research Program, 2012:29). The intelligence community is an obvious potential beneficiary of this effort.

> **Conclusion 4.3:** *Many of the scientific needs of the intelligence community regarding climate change adaptation and vulnerability are congruent with those of the USGCRP and various individual federal agencies. Intelligence agencies and the USGCRP can benefit by joining forces in appropriate ways to advance needed knowledge of vulnerability and adaptation to climate change and of the potential of climate change to create social and political stresses.*

A whole-of-government approach to understanding adaptation and vulnerability to climate change can advance the objectives of multiple agencies, avoid duplication of effort, and make better use of scarce resources. Such an interagency effort will help in anticipating the social and political consequences of climate events and in building the basis for a widely useful system for monitoring and analysis. This system would aid in anticipating security threats and could be employed by the U.S. intelligence community and other domestic and international entities to inform choices about responses to climate change.

Building Fundamental Understanding

> **Recommendations 3.1, 4.1, 5.1, and 6.1:** *The intelligence community should participate in a whole-of-government effort to inform choices about adapting to and reducing vulnerability to climate change.*

> **Recommendation 3.1:** *It should, along with appropriate federal science agencies, support research to improve the ability to quantify the likelihoods of potentially disruptive climate events, that is, single extreme climate events, event clusters, and event sequences. A special focus should be on quantifying risks of events and event clusters that could disrupt vital supply chains, such as for food grains or fuels, and thus contribute to global system shocks.*

This research should include efforts by climate scientists to improve fundamental understanding of the effects of climate change on the likelihoods of extreme climate events and also efforts to apply the methods of extreme value statistics to these problems, particularly the problem of estimating the likelihoods of clusters of extreme climate events that are dependent on the same underlying climatic processes. Such efforts would help in defining climate event scenarios for countries, regions, and systems that could be used as the basis for climate stress tests.

Recommendation 4.1: *It should, along with the USGCRP and other relevant science and mission agencies, develop priorities for research on climate vulnerability and adaptation and consider strategies for providing appropriate research support. The interagency effort on vulnerability and adaptation should include agencies responsible for community resilience and disaster preparedness and response domestically and internationally.*

Such an interagency process does not imply that climate change should be defined as a security issue. Rather, it indicates that security issues are among those that should be considered in developing and executing a research agenda on climate change adaptation and vulnerability.

Recommendation 5.1: *It should, along with other interested agencies, support research to improve understanding of the conditions under which climate-related natural disasters and disruptions of critical systems of life support do or do not lead to important security-relevant outcomes such as political instability, violent conflict, humanitarian disasters, and disruptive migration.*

Understanding the connections between harm suffered from climate events and political and social outcomes of security concern is arguably the most important aspect of climate change from a national security perspective, but it has received relatively little scientific attention until now. The disaster research community, which has been the locus of research on the political effects of climate events, has not been well connected to the climate research community.

To build the needed fundamental understanding will require the integration of knowledge of political and socioeconomic conditions in countries of interest; knowledge from climate science about the potential exposure of these countries to climate events; and knowledge from social science about the susceptibility of these countries to being harmed by those events and the likelihood of effective coping, response, and recovery at local to national levels. These sources of knowledge come from different communities of experts, which will need to communicate with each other but do not necessarily do so now. An important need is to integrate the social science of natural disasters and disaster response with other forms of analysis. This body of knowledge is particularly important for assessing the security consequences of climate change because disruptive climate events will typically be perceived and responded to as natural disasters. The recommended interagency process can help bring these communities of experts together, as they tend to associate with different groups of agencies.

Improving Monitoring and Analysis

Conclusion 6.1: *Monitoring to anticipate national security risks related to climate events should focus on five key types of phenomena:*

1. *Climate events and related biophysical environment phenomena;*
2. *The exposures of human populations and the systems that provide food, water, health, and other essentials to life and well-being;*
3. *The susceptibilities of people, assets, and resources to harm from climate events;*
4. *The ability to cope with, respond to, and recover from shocks; and*
5. *The potential for outcomes of inadequate coping, response, and recovery to rise to the level of concern for U.S. national security.*

Given that security threats arise from combinations of all of these phenomena, indicators and monitoring systems should be developed to follow them at various levels from local to national.

Conclusion 6.2: *Developing an adequate system for monitoring the conditions that can link climate events to national security concerns will require maintaining critical existing observational systems, programs, and databases; the collection of new data; the analysis of new and existing data; and the improvement of analytic systems, leading to better understanding of the linkages over time and to improved indicators of key variables where quantitative indicators are appropriate and feasible to produce. It will typically require finer-grained data than are currently available. It will also require improved techniques for integrating quantitative and qualitative information.*

We emphasize that improved understanding and monitoring of the various elements of climate vulnerability—a key link between climate events and security concerns—is an objective that the intelligence community shares with the USGCRP and many other institutions at federal, state, local, and international levels. To address the challenges of monitoring, which include both new and enduring methodological problems, the intelligence community needs to draw on knowledge from the academic research community, as some current efforts are already doing.

Recommendation 6.1: *One of the objectives of the recommended whole-of-government effort to inform choices about adapting to and reducing vulnerability to climate change should be to build the scientific basis for indicators in this domain.*

This effort would support activities by the research communities involved in assessing exposures and vulnerabilities to environmental change to identify a relatively small number of key variables relevant to the social and political consequences of climate events. The effort of the climate science community to identify a small number of "essential climate variables" suggests the kind of process that could be used.

Recommendation 6.2: *The U.S. government should begin immediately to develop a systematic and enduring whole-of-government strategy for monitoring threats connected to climate change. This strategy should be developed along with the development of priorities and support for research.*

The monitoring should include climate phenomena, exposures and vulnerabilities, and factors that might link aspects of climate and vulnerability to important security outcomes, and it should be applicable to climate issues globally. It should also include making and periodically updating priority judgments about when and where high-resolution monitoring is needed. Analysis will require the integration of quantitative indicators with traditional security and intelligence analytic methods.

The value of monitoring efforts is likely to increase over time because of improvements in monitoring systems and because potentially disruptive climate events are expected to increase in frequency and intensity in the future. Existing open-source monitoring systems that may provide useful information on key variables should be periodically examined for their potential utility, but with critical attention paid to indicator selection, data reliability and validity, and cross-case and cross-national comparability.

For the great majority of existing and potential indicators, the required spatial and temporal resolution is finer than what is currently available. High-resolution monitoring will be especially important for highly significant and highly vulnerable locations. The appropriate level of spatial and temporal resolution for indicators varies, however, with the substantive domain. In setting priorities for indicator development and improvement, the intelligence community should take into account the gaps between the existing and the desired resolution and should invest in improved resolution of those indicators judged to be the most needed and the most useful in places of concern.

It is important to develop and validate monitoring systems now in order to have baseline data for future studies of climate event impacts and for social and political stress analyses. Validation is particularly important for emerging monitoring technologies, such as those involving sophisticated data mining algorithms (e.g., of Internet postings) and remote observations that are overlaid on geographic information systems. Such techniques may produce outputs that catch the eye and are very impressive on first glance, but they are sometimes closely held by their developers and difficult to validate, especially if they involve infrequent events. Indicators and monitoring results should be interpreted with caution until these techniques develop a record of validation.

Organized international collaborations with potentially affected societies and governments and the open sharing of data will be important aspects of developing the needed monitoring systems. Such collaborations are likely to play a crucial role in gaining acceptance of higher-resolution monitoring at critically vulnerable locations. The collaborations are also likely to benefit many governments and international organizations that have a stake in reducing the risks of climate change to human and international security; the U.S. government in particular can benefit from data-gathering efforts in and by other countries. Of course, U.S. government agencies will continue to gather some kinds of information that will not be openly shared, and there will be questions about which data and information-gathering methods can and should be openly shared. Depending in part on how interagency collaborative relationships are structured and managed, there could also be suspicions related to the involvement of U.S. intelligence agencies in international information-gathering efforts related to security. Such issues will need to be addressed in ways that we have not had the opportunity to consider in this study. Nevertheless, the benefits of open, international data development and sharing should be taken seriously as work on monitoring systems proceeds. These benefits include the development of compatible concepts, databases, and indicators across countries, which help speed scientific progress and improve the ability to learn from experiences in other countries.

Improving the Capacity to Anticipate Security Threats

Recommendation 6.3: *The intelligence community should establish a system of periodic "stress testing" for countries, regions, and critical global systems regarding their ability to manage potentially disruptive climate events of concern. Stress tests would focus on potentially disruptive conjunctions of climate events and socioeconomic and political conditions.*

The intelligence community presumably already uses an analogous process to consider the ability of foreign governments and societies to withstand various kinds of social and political stresses. This recommendation calls on the community to incorporate climate risks and the associated exposures and vulnerabilities into such exercises. The concept of a climate stress test provides a framework for integrating climate and social variables more systematically and consistently within national security analysis.

A stress test is an exercise to assess the likely effects on particular countries, populations, or systems of potentially disruptive climate events. The recommended stress tests would involve analyzing the likely effects of an event at some projected time of occurrence in terms of key variables affecting susceptibility, coping, response, and recovery or the failure thereof, and the likely responses within regions or countries of interest in the event that these actions are perceived to be inadequate. The tests would draw on knowledge about the potential events and each of the other types of phenomena and would provide a major way of making knowledge about climate events, exposures, and vulnerabilities operational in security analysis.

Stress tests should assess the potential consequences for security of climate events under either of two conditions: when climate scientists can say with some confidence that the events will be increasingly likely to occur or become more severe, or when the events seem increasingly likely to occur based on a fundamental understanding of climate dynamics but available evidence is not yet sufficient for climate scientists to attach confidence to such projections. Stress tests might also be triggered by assessments indicating that event likelihood, exposure, or susceptibility is increasing or that the capacity to respond adequately to certain kinds of climate events is declining in a region or country of concern.

The results of stress tests would inform national security decision makers about places that are at risk of becoming security concerns as a result of climate events and could be used by the U.S. government or international aid agencies to target high-risk places for efforts to reduce susceptibilities or to improve coping, response, and recovery capacities. The stress testing process would also help advance understanding through an accumulation of data on potentially disruptive events and their social, political, and security consequences.

Countries, regions, and systems of particular security interest should be prime targets for periodic stress testing. Given the joint criteria of significant potential for climate change impacts and importance to U.S. national security, it is likely that no more than 12 to 15 countries will need to be monitored and subjected to periodic stress tests over the next decade, many of which are likely to be in critical, and often shared, watershed areas in South Asia, the Middle East, and Africa. If the criteria for importance to the United States are expanded to include foreign policy and humanitarian concerns, then the number of countries to be monitored and stress-tested regularly over the next decade may rise to between 50 and 60. Stress testing should also be applied periodically to global systems that meet critical needs, including food supply systems, global public health systems, supply chains for critical materials, and disaster relief systems.

Decision science techniques should be used and further developed to ensure that the stress tests make the best use of the available information. Stress testing might draw on various methods, including the qualitative interpretation of available knowledge, formal modeling, and interactive gaming approaches. Decision science techniques should be employed to design the processes and interpret the input from different kinds of expertise and modes of analysis in order to make the best possible use of information. The stress-testing exercises should themselves be monitored and critically evaluated so that stress-testing methods can be improved over time.

1

Climate Change as a National Security Concern

Over the past several years the U.S. intelligence community has engaged with the National Academy of Sciences and National Research Council (NRC) concerning a range of issues related to climate and security. A standing Committee on Climate, Energy, and National Security (CENS) was established in 2008 to facilitate the increased involvement of scientists in answering questions related to climate and environmental change, energy, natural disasters, and security. The committee undertakes activities to bring scientific expertise to bear on questions of importance to the intelligence community related to climate and environmental change (see, for example, National Research Council, 2010b, 2010d, 2012b). The CENS activities led to a request in 2010 for a study to, among other tasks, "identify ways to increase the ability of the intelligence community to take climate change into account in assessing political and social stresses with implications for U.S. national security." The complete statement of task for this study appears in Box 1-1.

To carry out this task the NRC created a committee with broad expertise in the physical and social sciences and in security matters. The goal was for the committee to be able to integrate knowledge from across the physical and social sciences and also to be able to offer advice to the intelligence community on how to think about the security risks that might arise when climate change leads to situations for which countries, regions, or human life-supporting systems are not adequately prepared. Biographies of the committee members may be found in Appendix A.

This study focuses on some of the ways that climate change might create or alter risks to U.S. national security, in particular, ways that fall

> **BOX 1-1**
> **Statement of Task: Assessing the Impact of Climate Change on Social and Political Stresses**
>
> The National Research Council (NRC) would undertake a study to evaluate the evidence on possible connections between climate change and U.S. national security concerns and to identify ways to increase the ability of the intelligence community to take climate change into account in assessing political and social stresses with implications for U.S. national security. The study panel would focus on several broad questions, such as: What are the major social and political factors affecting the relationship between climate change and outcomes relevant to U.S. national security? What is the basis for this knowledge and how strong is it? What research and measurement strategies would strengthen the basis for this knowledge?
>
> The study panel would develop a conceptual framework for addressing such issues on the basis of two workshops, existing research literature, and relevant NRC studies. It would produce a report including its conceptual framework and findings and conclusions regarding the key climate-security connections and issues of assessment of climate-related security risks examined in the workshops and the scientific literature. It would also identify variables that should be monitored and ways that indicators of climate change, impacts, and vulnerabilities might be developed and made useful to the intelligence community in assessing climate-related threats to U.S. national security.

within the mission of the U.S. intelligence community. This mission covers a broad range of risks. It includes possible military attacks on the United States, its allies and partners, and American facilities overseas, but it is much broader. The intelligence community is also responsible for assessing the likelihood of violent subnational conflicts in countries and regions with extremist groups, dangerous weapons, critical resources, or other conditions of security concern. It must also anticipate and assess various other risks to the stability of states and regions and risks of major humanitarian disasters in key regions of the world, both because of the indirect threats such risks may pose to the United States or its allies and because of national commitment to the principles of U.N. Security Council Resolution 1674, which proclaims "the responsibility to protect populations from genocide, war crimes, ethnic cleansing and crimes against humanity."

Given these intelligence mission elements, the central questions motivating this study are: How might climate change lead to new or increased risks to U.S. national security? Might it, for example, put new stresses on societies or on systems that support human well-being, such as supply chains for food or energy, and thus pose or alter security risks to the United States? Will intelligence and security organizations need to gather new kinds of information or synthesize existing information in new ways in order to assess climate-related security risks? Will they need to develop new ways to anticipate and assess security risks to address those that are affected by climate change?

This report is based on current understanding of the state of the climate system as assessed internationally by the Intergovernmental Panel on Climate Change (e.g., Intergovernmental Panel on Climate Change, 2007, 2012), as assessed nationally in reports by the U.S. Global Change Research Program (USGCRP) and by the NRC within the suite of congressionally mandated studies known as America's Climate Choices (National Research Council, 2010a, 2010b, 2010c, 2010d) and in subsequent relevant reports (e.g., National Research Council, 2011a, 2011b, 2012b), and in other relevant literature reviewed by the committee. The committee's purpose was not to readdress the science of climate change or to review past assessments, but to build on this knowledge to address the issues in the statement of task.

POTENTIAL CLIMATE–SECURITY CONNECTIONS

Over the past decade, several groups within the U.S. security policy community, both within and outside government, have given increasing attention to the potential risks that climate change could pose for national as well as international security. In 2008, for example, the intelligence community produced *The National Intelligence Assessment on the National Security Implications of Global Climate Change to 2030* (Fingar, 2008).[1] Climate issues were included in the 2010 *Quadrennial Defense Review* (*QDR*) (U.S. Department of Defense, 2010) as well as in the 2010 edition of the *National Security Strategy* (White House, 2010).

In addition to the attention from the U.S. government, beginning in the mid-2000s many foreign and security policy think tanks and research organizations produced reports on the potential connections between climate change and security. The reports were generally the work of groups of security experts, informed by consultations with climate scientists and regional and country specialists. Some reports also examined evidence from the social sciences. The groups drew upon this collective expertise to project a range of scenarios for potential impacts, usually over a 20-year period,

---

[1]The assessment itself is still classified, but the methodology and principal conclusions of the report were presented in the statement for the record prepared in conjunction with testimony to the House Permanent Select Committee on Intelligence and the House Select Committee on Energy Independence and Global Warming. The National Intelligence Council also sponsored an extensive set of unclassified reports and conferences on the potential effects of climate change on key regions and countries; the materials may be found at http://www.dni.gov/index.php/about/organization/national-intelligence-council-nic-publications (accessed September 27, 2012).

although some also included projections to the end of the century. Given that these were not academic reports, the basis for the groups' judgments and the level of confidence associated with them were usually not specified in detail. Without attempting a comprehensive review, this section seeks to provide a summary of frequently occurring themes and arguments about climate–security connections from major government reports (Fingar, 2008; U.S. Department of Defense, 2010; White House, 2010; Defense Science Board, 2011) and from some of the best-known examples in the mainstream policy literature (Busby, 2007; Center for Naval Analysis, 2007; Lennon et al., 2007; Center for Climate and Energy Solutions, 2009; Carmen et al., 2010; International Institute for Strategic Studies, 2011; Treverton et al., 2012). For some key statements from these studies, see Box 1-2.

These government and policy documents reflect a number of important common elements. Above all, the connections between climate and security are not presumed to be direct; they are seen as complicated and contingent, with the effects of climate events felt through their consequences for other factors that then affect security. For example, the 2010 *QDR* concludes:

> While climate change alone does not cause conflict, it may act as an accelerant of instability or conflict, placing a burden to respond on civilian institutions and militaries around the world. In addition, extreme weather events may lead to increased demands for defense support to civil authorities for humanitarian assistance or disaster response both within the United States and overseas. (U.S. Department of Defense, 2010:85)

The most frequently cited potential climate events include sea-level rise, the shrinking of glaciers and the Arctic icecap, an increase in extreme weather events, and increasingly intense droughts, floods, and heat waves. The scenarios and examples presented in the above reports address broad consequences for fundamental societal needs such as food, health, and water and also the likely implications for specific regions and countries. Although the reports generally agree that future climate events are likely to increase tensions and political instability within and between states and perhaps also increase internal conflicts, they do not forecast an increase in interstate conflict.

Taken together, the most commonly cited climate–security scenarios in these reports result from failures or shortcomings of human systems in adapting to a changing climate; that is, they turn on the vulnerabilities of these systems to climate events. In these scenarios climate events cause harm to various support systems for human life and well-being by exceeding the ability of these systems to cope. Depending on other social, economic, political, and environmental factors, the harm may result in larger-scale political and social outcomes that are of concern for U.S. national security. All of the reports include some scenarios of this sort, although different re-

BOX 1-2
Statements About Climate and Security
Connections from Previous Security Analysis

"Climate change acts as a threat multiplier for instability in some of the most volatile regions of the world." (Center for Naval Analysis, 2007:6)

"[T]he United States can expect that climate change will exacerbate already existing north–south tensions, dramatically increase global migration both inside and between nations (including into the United States), spur more serious public health problems, heighten interstate tension and possibly conflict over resources, challenge the institutions of global governance, cause potentially destabilizing domestic political and social repercussions, and stir unpredictable shifts in the global balance of power, particularly where China is concerned. The state of humanity could be altered in ways that create strong moral dilemmas for those charged with wielding national power, and also in ways that may either erode or enhance America's place in the world." (Lennon et al., 2007:103)

"We assess that climate change alone is unlikely to trigger state failure in any state out to 2030, but the impacts will worsen existing problems—such as poverty, social tensions, environmental degradation, ineffectual leadership, and weak political institutions. Climate change could threaten domestic stability in some states, potentially contributing to intra- or, less likely, interstate conflict, particularly over access to increasingly scarce water resources." (Fingar, 2008:4–5)

"Since climate change affects the distribution and availability of critical natural resources, it can act as a 'threat multiplier' by causing mass migrations and exacerbating conditions that can lead to social unrest and armed conflict." (Center for Climate and Energy Solutions, 2009:1)

"While climate change alone does not cause conflict, it may act as an accelerant of instability or conflict, placing a burden to respond on civilian institutions and militaries around the world. In addition, extreme weather events may lead to increased demands for defense support to civil authorities for humanitarian assistance or disaster response both within the United States and overseas." (U.S. Department of Defense, 2010:85)

"Climate change is likely to have the greatest impact on security through its indirect effects on conflict and vulnerability." (Defense Science Board, 2011:xi)

"Climate change is not happening in a vacuum: in many areas of the world it will be accompanied by rapid population growth, resource shortages, and energy price increases. Analytically, it is difficult to separate the effects of climate change from other factors, such as food shortages, migration, ethnic tensions and other issues that could drive violence. However, the potential impacts of climate change on water, energy, and agriculture will make it a central driver of conflict. The impacts of climate change combine to make it a clear threat to collective security and global order in the first half of the 21st Century." (International Institute for Strategic Studies, 2011:11) ports emphasize the effects of climate change on different support systems. Declines in food and water security are among the most frequently cited kinds of harm (e.g., Busby, 2007; Center for Naval Analysis, 2007; Lennon et al., 2007; Fingar, 2008; Center for Climate and Energy Solutions, 2009; Carmen et al., 2010; U.S. Department of Defense, 2010; Defense Science Board, 2011; International Institute for Strategic Studies, 2011; Treverton et al., 2012), and sub-Saharan Africa is often singled out as the region most likely to experience the greatest effects on security. For example, Fingar (2008) wrote:

> We judge that sub-Saharan Africa will continue to be the most vulnerable region to climate change because of multiple environmental, economic, political, and social stresses. . . . Many African countries already challenged by persistent poverty, frequent natural disasters, weak governance, and high dependence on agriculture probably will face a significantly higher exposure to water stress owing to climate change. (p. 8)

In some of the scenarios increasing food and water insecurity interact to increase risks to health (e.g., Busby, 2007; Center for Naval Analysis, 2007; Lennon et al., 2007). In others health risks result from changes in weather patterns that shift the ranges for vector-borne diseases (Center for Naval Analysis, 2007; Lennon et al., 2007). Several scenarios see such declines in food or water security or disease outbreaks as likely drivers of population migrations, both within and across borders, that result in political or social stress, usually in the countries that receive the immigrant populations (e.g., Busby, 2007; Center for Naval Analysis, 2007; Lennon et al., 2007; Fingar, 2008; Center for Climate and Energy Solutions, 2009; Treverton et al., 2012). Two of the most-often cited scenarios are increased flooding or a rise in sea level forcing millions of Bangladeshis into India and an increasing desertification and drought forcing people from northern and sub-Saharan Africa into Europe. In both scenarios immigration issues are already a source of major tension. Energy security also figures prominently in several projected climate–security scenarios (Lennon et al., 2007; U.S. Department of Defense, 2010; International Institute for Strategic Studies, 2011), in which climate change is seen not only as yielding potential benefits for natural gas and perhaps biofuels producers but also as increasing the vulnerability of countries and industrial systems that rely on imported fuel (Lennon et al., 2007).

The paths envisioned from climate events to specific security consequences are often complicated. For example, tensions could increase over access to increasingly scarce resources, and that escalation, especially if it led to overt conflict, could in turn further limit access to resources so that people who had not previously been affected would now face shortages. Some scenarios suggest that diminished national capacity or outright state failure would create increasing opportunities for extremism or terrorism. Again, sub-Saharan Africa is often cited as the most vulnerable region.

In addition to these specific scenarios, many of the reports foresee increasingly frequent and increasingly severe natural disasters that will strain the capacity to cope with the resulting humanitarian emergencies, both in the United States and overseas (Busby, 2007; Center for Naval Analysis, 2007; Center for Climate and Energy Solutions, 2009). This is of particular concern to the U.S. military, given its expanding role in disaster assistance, although several of the reports note that helping countries prepare for and cope with disasters offers important opportunities for positive engagement.

These climate–security analyses raise concerns about several security issues beyond those of inadequate adaptation leading to humanitarian disasters, political instability, or violent conflict. One class of scenarios involves direct threats of climate change to the ability of the U.S. military to conduct its missions. An example is the threat that sea level rise, possibly in combination with more intense coastal storms, poses to naval bases in low-lying coastal areas (Busby, 2007; Center for Naval Analysis, 2007; Center for Climate and Energy Solutions, 2009; U.S. Department of Defense, 2010).[2] More generally, analyses foresee climate change having broad negative effects on military organization, training, and operations—for example, by exacerbating operational difficulties for troops and equipment in already difficult locations (Busby, 2007; Center for Naval Analysis, 2007; Center for Climate and Energy Solutions, 2009; Carmen et al., 2010; U.S. Department of Defense, 2010; National Research Council, 2011b). Other concerns include the vulnerability of U.S. Department of Defense (DOD) fuel supplies to severe weather that disrupts supply lines and the possibility of droughts restricting access to water for forces and facilities overseas.

Perhaps the most frequently cited security risk from climate change is the possibility of melting Arctic sea ice leading to increased international tensions over newly accessible sea routes and natural resources in the Arctic (Busby, 2007; Center for Naval Analysis, 2007; Carmen et al., 2010). A recent NRC study (National Research Council, 2011b), addresses these and other security issues of interest to the U.S. naval forces.

INCREASING RISKS OF DISRUPTIVE CLIMATE EVENTS

It is now clear from an accumulation of scientific evidence that the risks of potentially disruptive climate events are increasing. The scientific evidence on this point is aptly summarized in this conclusion from a recent major review of the science by the NRC: "Climate change is occurring, is

---

[2] For examples of the severe damage suffered by U.S. bases in the past from hurricanes, see Busby (2007:6).

caused largely by human activities, and poses significant risks for—and in many cases is already affecting—a broad range of human and natural systems" (National Research Council, 2010a:3). These increased risks will not be reduced anytime soon: "Human-induced climate change and its impacts will continue for many decades and in some cases for many centuries. Individually and collectively, these changes pose risks for a wide range of human and environmental systems, including . . . national security" (National Research Council, 2010a:4). Moreover, many of the forces driving anthropogenic climate change, chiefly emissions of carbon dioxide and other "greenhouse gases" and changes in land use that increase net absorption of energy from the sun, have been increasing at an accelerating rate over the past century or so (National Research Council, 2010a). Accordingly, global average temperatures have been increasing substantially over the past century (NASA Goddard Institute for Space Studies, 2012), and, given the long residence time of carbon dioxide within the atmosphere, further temperature increases are projected for at least the rest of the current century even under scenarios in which past emissions trends are significantly curtailed (Meehl et al., 2007). As another recent NRC report pointed out, "Recently experienced climatic events are not likely to serve as guides to what to expect next" (National Research Council, 2009:14).

In short, it is becoming increasingly likely that the world will experience climate-related conditions it has not seen before. The frequency of natural disasters related to weather and climate has been increasing for at least three decades, as have losses from these events (Intergovernmental Panel on Climate Change, 2012; Munich Re, 2012). Temperature trends at the local level show both increasing average temperatures and increasingly frequent occurrences of high temperatures that were quite rare in the 1951–1981 period (see Figure 1-1). As discussed further in Chapter 3, temperature increases have implications for the hydrological cycle because for each 1°C in global mean surface temperature there is a corresponding 7 percent increase in atmospheric water vapor.

These trends indicate that high-temperature extremes are becoming more common even more rapidly than the average temperature is increasing and that the rate of change is increasing. Such trends in extreme events and the current understanding of climate change provide ample reason to expect these weather and climate trends to continue, along with the considerable spatial and interannual variability that has been experienced in the recent past. Presently, the ratio of record daily high temperatures to record low temperatures at U.S. observing stations is 2 to 1, rather than the 1:1 ratio that would be expected if climate were not warming. For a mid-range emission scenario, the ratio has been projected to increase to 20 to 1 by mid-century and to roughly 50 to 1 by the end of the century (Meehl et al., 2009).

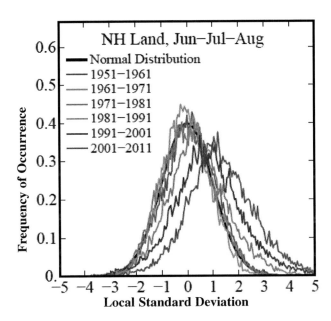

FIGURE 1-1 Frequency of occurrence of local temperature anomalies on the Northern Hemisphere (NH) land surface by decades, in units of the local standard deviation. Anomalies are relative to the 1951–1980 base period. Compared to 1951–1961, the distribution of anomalies has steadily moved in the direction of warmer temperatures in the 1980s, 1990s, and 2000s, with warm extremes becoming more common in each of these decades. Occurrences of summer temperatures more than three standard deviations above the mean, which were extremely rare before 1980, occurred an order of magnitude more frequently in the 2000s and covered between 4 percent and 13 percent of the world between 2006 and 2011.
SOURCE: Hansen et al. (2012).

The 2009 NRC report appears to have been prescient in making this observation: "[D]ecision makers must expect to be surprised—probably with increasing frequency" (National Research Council, 2009:18). An important reason to expect to be surprised is that Earth's climate is changing at a rate that is unprecedented, at least throughout human history. The rate of carbon dioxide buildup in the atmosphere is now a factor of 10,000 greater than it was during any period on geological record prior to human civilization, and sea levels during prior interglacial periods with comparable average surface temperatures were substantially higher than they currently are (Hansen et al., 2010). The unprecedented rate of carbon dioxide accumulation means that Earth's climate system—and likely its ecological system as well—will continue to undergo a very large energy balance adjustment, possibly at an unprecedented rate. One can confidently expect that there will be significant consequences. Although we do not know the exact magnitude, timing, or character of all of these consequences, it is prudent to assume that some of them will appear as surprises in the form of unanticipated events that compel some reaction.

National security decision makers do not like surprises and expect the intelligence community to provide sufficient warning to make it possible to avoid, ameliorate, or alter the undesired consequences of emerging developments. Fundamental knowledge of climate dynamics indicates that many types of extreme climate events are likely to become more frequent, even though we do not know enough to predict which extreme events will occur where or when. Although it is true that old climate averages are not a good guide to the future, recent experiences of climate-driven surprise *are* likely to be a good guide to what to expect next. In Chapter 3 we discuss in more detail what climate science can and cannot tell us about what events to expect in the future.

From the standpoint of those who must deal with the consequences of any specific climate-related event, it may make little difference whether the event can be reliably attributed to anthropogenic climate change or whether it instead results from natural climate variation. But from the standpoint of anyone concerned with the global profile of security risks, it is important to recognize that such events are expected to become more common because of anthropogenic climate change. Whether any specific event can be attributed to anthropogenic climate change is less relevant than the likelihood of serious climate-related disruptions occurring in places where they might raise security risks for the United States and the change in that likelihood over time. Security analysis must be based on an accurate assessment of the risks, and the risks from climate events are changing. It is also worth noting that many other countries are contemplating—or taking, or want to take—steps to address the emerging risks of climate change. This provides a basis for cooperative action on reducing these risks as well as a reason to regularly reassess vulnerabilities to the effects of climate change.

What "Climate Change" Means in This Study

Climate change is commonly defined by climate scientists as change in the mean or variability of a climate property that persists for an extended period, typically a decade or longer (Intergovernmental Panel on Climate Change, 2012). In this study we are interested in both climate change and climate variability. Climate variability refers to variations from the mean (and other statistical properties) of the climate at all spatial and temporal scales beyond that of individual weather events. (Definitions adapted from

Intergovernmental Panel on Climate Change, 2012.) Among the various types of climate variability we are particularly interested in those variations, such as the El Niño–Southern Oscillation, the North Atlantic Oscillation, and others, that have effects for multiple years and for which early warning is possible.

We include both climate change and climate variability because it will generally be climate events, particularly events now considered extreme, which will lead to various security-related issues. Normal climate variability has in the past led to a few such extreme climate events, but these events can be expected to become more frequent with climate change. As we discuss in Chapter 3, the combination of climate change and climate variability will further alter the frequency of such events. These climate events can be politically and socially disruptive and may require U.S. government action independent of whether or not particular events can confidently be attributed to anthropogenic climate change.

Thinking About Unlikely Extreme Events

Occasions in which climate change may contribute to events requiring U.S. government action may arise in the context of climate surprises or as the result of the conjunction of climate events and social or political developments. The contributing climate events are typically extreme events and therefore, by definition, have been highly unlikely in the past. They may either be acute (e.g., a storm) or slowly developing (e.g., a drought). After such events occur it is often easy to identify the relevant precursors, but before the event it is often much more difficult to separate the signal from the noise. When considering the relationships of climate change to national security before the occurrence of such events, intelligence analysts need ways to decide which of the many unlikely events that could contribute to a crisis are most worthy of attention.

It is important to recognize that climate scientists and security policy makers tend to have very different ways of thinking about unlikely events. This is particularly the case with extreme climate events, which have been quite infrequent but can also be highly consequential. Although extreme events have the greatest potential to disrupt political and social systems and raise security concerns and are thus the most important events for security analysts to understand, their low frequency of occurrence makes it especially difficult to validate scientific predictions about them.

Scientists tend to be conservative in making claims about the future because of their usual rules of inference. In making projections about the climate future, especially of unlikely events, scientists develop and validate complex models and test them by attempting to reproduce documented climate trends and events. They typically initialize models with observational data beginning at some point in the past and test retrospective model predictions (i.e., hindcasts) against known observations between the initial point and the present. To the extent that understanding can be demonstrated by this method, they then project future climate based on assumptions of future greenhouse gas concentrations that are predicated on scenarios of economic growth, energy consumption, and demographic change, carefully noting the uncertainties in the projections. If projections using several different models tend to agree, scientists' confidence in the results increases.

This approach has certain limitations, particularly in dealing with highly infrequent and unprecedented events. For such events there are relatively few data with which to build the models and test them against past experience, and, as a result, models that make slightly different assumptions can lead to different projections of the future likelihoods of those events. When this happens, scientists' confidence in projections is low, and they are reluctant to make statements about the future likelihood of such events. This may not mean that such events can safely be considered to be highly unlikely, but only that there are not enough data to make a confident projection of likelihood.

It is in the nature of the scientific enterprise that claims are not made about the existence of phenomena unless a high level of confidence can be asserted for such claims. A 95 percent confidence level is a typical cut-off point: Scientists defer making claims or predictions until they can be highly confident in them. In short, science requires a high burden of proof for concluding that a climate phenomenon exists or that a projection of future climate conditions—especially extreme conditions—is solidly supported.

In security policy the practice for deciding whether to take a hazard seriously is much different from the practice in making scientific claims. Security analysts are focused on risk, which is usually understood to be the likelihood of an event multiplied by the seriousness of the consequences if it should occur. Thus security analysts become seriously concerned about very high-consequence negative events even if scientists cannot estimate the probability of their occurrence with confidence and, indeed, sometimes even if they are fairly confident that the probability is quite low. During the Cold War, for example, most people thought that deterrence was robust, and few thought the likelihood that the Soviet Union would actually initiate a nuclear attack against the United States was anything but minuscule. But because the consequences would have been so dire, tremendous efforts were made by the intelligence and national security communities to monitor events that might provide early warning of the possibility of such a strike. The same is true of threats of terrorist attacks on the U.S. homeland today. Even though there have been few terrorist attacks altogether—and no major ones on the United States since 2001—substantial resources are allocated to identifying threats and reducing risks. The public and elected officials have consistently supported such risk-based allocations of resources.

Unlikely Extreme Events in an Unprecedented Climate

From the perspective of security, events that could disrupt the social and political systems of importance to U.S. national security are particularly important. Certain climate events—most likely, rare and extreme events—could meet this criterion. These are, by definition, events on the long tails of the probability distributions of climate events. Unfortunately, the climate science community is much less confident in its skill at projecting the rate of change in the frequency or magnitude of events at the extremes of such distributions than it is in projecting averages or even the likelihoods of events in the middle 80 to 90 percent of the distributions. One reason for caution is, as already noted, that there is a very limited population of such events to use for validating models. Moreover, the spatial resolution of climate change models is often coarser than is required to resolve the spatial structure of many extreme events to the degree needed for security analysis.

Another factor limiting confidence in the projections of extreme climate events is that the fundamental attributes of Earth's climate system have moved or very soon will move beyond the bounds of experience on which models are based. For example, the concentrations of greenhouse gases (GHGs) in the atmosphere are now greater than they have been for at least 800,000 years (National Research Council, 2010a), and the current rate of carbon dioxide accumulation in the atmosphere is at least an order of magnitude greater than the natural rate that prevailed prior to the rise of human civilizations (see http://www.ncdc.noaa.gov/paleo/icecore/antarctica/vostok/vostok_co2.html [accessed November 14, 2012]). As climate moves outside the range of experience, models of the effects of higher GHG concentrations cannot be validated against the kinds of high-resolution observational data that provide the most desirable basis for model testing. Global average temperature already is or soon will be higher than it has been at any time in recorded human history, and it is increasing at an unprecedented rate (Intergovernmental Panel on Climate Change, 2007; National Research Council, 2009). Moreover, the variance in temperature indicators has been increasing over the past half century (see Figure 1-1). All of these phenomena—higher temperatures, higher variation in temperatures, and rapid change—increase the likelihood of the occurrence of extreme temperature events. In addition, as the climate moves outside the bounds of the experience on which existing models are validated regarding the averages and ranges of variation of climate parameters and their rates of change, scientists may attach less confidence to projected extremes simulated by the models.

Unprecedented rates of change in the global average temperature and in the GHG concentrations that drive this change are particularly challenging for the climate modeling enterprise because of the well-known fundamental properties of complex systems, of which Earth's climate is a paradigmatic example. Because of the state and rate of change of Earth's climate over recent decades, confident projections of extreme events are especially difficult to produce. This does not mean that climate science has nothing to say about the future of extreme events that can be useful to the intelligence community. What it means is that there are multiple scenarios of the future of climate events that are each likely enough that they deserve consideration by the intelligence community. They should not be treated as predictions but rather as possibilities for evaluation in terms of the social and political scenarios they might set in motion, the security issues that might ensue, and the preparedness of the U.S. government to deal with the consequences.

Fundamental climate science provides some useful concepts for thinking about the future of climate events despite the limits of predictability of particular events. Consider, for example, the implications of the fact that although Earth's temperature remains well above the long-term average, the decade beginning in 1998 represented a hiatus in the longer overall global warming trend (National Aeronautics and Space Administration, 2012). A fundamental understanding of Earth's climate system makes it clear that global warming has not stopped and that the hiatus will be brief. The past two years suggest that it may already have ended (Foster and Rahmstorf, 2011; National Aeronautics and Space Administration, 2012).

The past 130 years or so include periods with strong warming and periods with little or no warming (e.g., Easterling and Wehner, 2009). As we discuss further in Chapter 3, even under continuing climate change decades with no warming can be expected in the future—along with decades with above average rates of warming. The coupled climate system has naturally occurring decadal signals, such as the Pacific Decadal Oscillation and the Atlantic Multi-Decadal Oscillation, which can serve to mask or accelerate the average rate of warming on decadal time-scales. The underlying trends can be understood in relation to fundamental processes of energy balance: If the climate system is continuing to absorb more energy from the sun than it is emitting, as must happen when greenhouse gas concentrations are increasing, then that energy remains in the Earth system and must show itself sooner or later through increased temperatures along with other changes in the climate system. Recent research (Meehl et al., 2011) suggests that during the hiatus in recent years most of this excess energy has gone into the deep oceans, where it may show itself through the enhancement of coupled oceanic-atmospheric phenomena, such as a strong El Niño and warming ocean temperatures.

Analyzing Plausible Extreme Event Scenarios

Climate scientists have paid significant attention to some of the ways that gradual global climate change might lead to abrupt and sometimes irreversible large changes at a continental or regional scale (National Research Council, 2002; Alley et al., 2003; Lenton et al., 2008). None of these possibilities can be projected with a level of confidence that would satisfy climate scientists. However, from a security analysis standpoint, in which attention is paid to future scenarios that are of sufficiently high consequence even if their probabilities are relatively low, this scientific work points to particular possible extreme event scenarios that are worthy of further analysis.

An example of the kind of process that could lead to surprising and very extreme events can be drawn from evidence in the paleoclimate records combined with recognition of enhanced polar temperature variations due to changes in GHG concentrations. Citing an observation by Bintanja et al. (2005) that over the past 800,000 years a 1°C increase in global mean temperature was associated with increased equilibrium sea levels of about 20 meters, Hansen and Sato (2012) have suggested that the sea level rise in the next century may well be on the order of 5 meters. They argue that an increase of 3.6°F (2°C) over pre-industrial temperature levels, which is highly likely to occur in this century, would commit the planet to sea level rise of many meters. Given the considerable uncertainty in the science of glaciology about the stability of major ice sheets, it is unclear whether their contribution to sea level rise over the next century will be linear or will follow a nonlinear trajectory with an increasing rate of change over time. If nonlinear processes prevail, then the common projection of up to 1 meter by the end of the century may be a lower bound rather than an upper bound. The rate at which the sea level rise would occur is critically important, of course, in terms of the social and political consequences.

To better evaluate the import for U.S. national security of scenarios like this, which have some scientific plausibility but which extend beyond the current scientific consensus, the intelligence community might benefit from several types of knowledge that could be developed in the coming decade to help analysts anticipate security issues that might arise if such a scenario is realized. These would include improved measures of rates of change in temperature and glacier ice cover in the polar regions; the use of existing climate models to project how this degree of ice melting would affect such outcomes as coastal inundation, extreme precipitation, and cyclonic storm severity; and assessments of the exposure, vulnerability, and response capacity of key countries and regions to these outcomes.

Several other examples of potential rapid-onset extreme climate event scenarios can readily be found (Lenton et al., 2008). For instance, models of changes in the Indian summer monsoon indicate that several sharply different but potentially dangerous shifts in the intensity of the monsoon are plausible, with the changes possibly occurring with a transition time of only a year or so. From a security perspective it may make sense to take each of the model-projected futures through a what-if scenario mode. Similarly, projections of the West African monsoon point to a Sahel (the east–west stretch of Africa south of the Sahara desert and north of the Sudanian savannahs) that is either wetter or drier or else has no average change in rainfall but has a doubling of the number of anomalously dry years (Lenton et al., 2008)—three scenarios that could be examined in terms of their social and political implications.

THE FOCUS OF THIS STUDY

The purpose of this study is to help improve the ability of the U.S. intelligence community and other interested actors to foresee security risks that may arise from climate change and its interactions with other social, economic, and political processes. Thus, we are concerned with climate risks to the extent that they may affect security risks.

Improved foresight can inform several kinds of policy responses: (1) responses to reduce climate risks (i.e., the risks that potentially harmful climate events will occur); (2) responses to reduce the exposure of people or valued assets to potentially harmful climate events; (3) responses to reduce susceptibility to harm from such events; and (4) responses that assist in the coping, response, and recovery processes after harmful events occur. We are aware of debates about how such policy responses should be organized and by whom. These questions are beyond the scope of our study, as are questions about how best to reduce the risks of occurrence of harmful climate events. Our focus is on anticipating security risks related to climate processes, understanding the roles of climatic and other factors in the dynamics of these risks, and informing decision makers about the nature of these risks and the opportunities for reducing them. We hope the study will, by improving understanding of the risks, provide a better basis for informed debate about which policy responses are most advisable. We have focused the study in three important ways, as outlined below.

Focus on Vulnerability to Climate Events

Although each of the scenario types that have been mentioned in climate–security studies is potentially significant for national security, this study focuses on scenarios involving the vulnerability of human populations, institutions, and life-supporting systems to harm from climate events, in which the harm has the potential to set events in motion that lead to security concerns. As discussed above, these scenarios in which climate events cause sufficient harm to human well-being to create humanitarian crises, political violence, or other issues of security concern are among the most prominently cited in the reports on climate change and security that have appeared in recent years from U.S. government security agencies and the foreign and security policy community. Such vulnerability-based scenarios predominated among the sponsor's concerns in requesting this study and in the early deliberations of the committee in open session with the sponsor present. The committee decided at the outset that these concerns provided the most appropriate focus for its work, as is discussed in further detail when the conceptual framework is presented in Chapter 2.

We acknowledge that with this focus, this study sets aside some climate–security connections that could prove highly significant and that deserve further study and analysis. These include some potential threats already noted, such as from extreme climate events that may impede the ability of U.S. military organizations to perform their missions and from conflict over natural resources and sea lanes in Arctic regions that may become newly accessible as a result of the melting of sea ice (cf., National Research Council, 2011b).

Another important class of security risks that are excluded from this study is those that may arise from policy responses to the anticipation or experience of disruptive climate events. Several plausible security risk scenarios begin with policies to limit climate change. For example, the expanded use of nuclear power in some countries to replace fossil fuels could increase risks of nuclear proliferation. Some policies to increase biofuel production could contribute to food price spikes and thus reduce effective food availability to low-income populations around the world. A single country's decision to counter global warming by geoengineering, perhaps by fertilizing the ocean to grow photosynthetic organisms or by injecting sulfate particles into the stratosphere, could create conflict with other countries. Several other policy-based scenarios begin with a country's efforts to protect itself from the expected consequences of climate change in ways that could disrupt international relations. For example, an upstream country might impound water from a river to guard against drought and thus reduce water supplies for its downstream neighbors. Or one country might purchase land in another country to produce food for its domestic consumption, creating conflict if a future food shortage hits the country where the food is being produced for export.

A number of threat scenarios of the above types are mentioned in previous climate–security analyses. Although some of them could have significant security consequences, they have not been treated as primary concerns in these reports. We have focused more narrowly on situations in which direct harm from climate events affecting vulnerable places or critical life-supporting systems could play a driving role in events of security concern. Policy responses intended to reduce one nation's vulnerability to climate events may sometimes increase the vulnerability of other countries. These kinds of climate-security connections could prove highly significant and deserve serious attention in security analysis, both by monitoring the development of such policies and by analyzing their implications for stresses in other places both when they are put in place and when a stressful climate event subsequently occurs.

Focus on Disruptions Outside the United States

Our study focuses largely on developments and vulnerabilities external to the United States, while recognizing that climate change is a global phenomenon and that events occurring within the United States can be disruptive in other countries, and vice versa. We examine some of these connections but not others. For example, a drought in U.S. agricultural areas that led to a spike in the global price of corn or wheat could lead indirectly to a humanitarian or political crisis elsewhere that could become a national security issue for the United States. Our study does examine such scenarios, but it does not examine the social and political consequences such events might have within the United States, nor does it examine the social and political consequences within the United States of climate events occurring elsewhere that disrupt global systems such as public health or the supply systems for critical commodities.

We emphasize, however, that such a separation between domestic and foreign impacts reflects only the division of missions among federal agencies, not the characteristics of climate phenomena or their consequences. In particular, observations, analyses, and fundamental knowledge that need to be developed in order to understand changing vulnerabilities to harm from climate events, which can offer valuable information to the U.S. intelligence community, are equally important for informing other federal agencies and decision makers below the national level, particularly including agencies responsible for domestic security and disaster management. They are also critical for informing international organizations. In Chapter 6 we discuss the needs for monitoring and analysis within a whole-of-government approach to developing an understanding of the effects of the changing risks of climate events.

Focus on the Next Decade

Given the risks and difficulties of projecting political, economic, and technical developments more than a decade into the future and the fact that countries are—and should be—starting now to contemplate steps to reduce vulnerability to climate change effects, this study focuses primarily on the next decade. In this way the study differs from most past analyses of climate change and its security implications. We consider policy and intelligence-gathering actions related to events that might occur in the coming decade as well as activities that must begin within a decade in order to have adequate intelligence capacity for anticipating climate–security interactions at later times. An adequate intelligence capacity in this area must include an improved ability to anticipate changes in climate-related security risks beyond the decadal time horizon, for at least two reasons: The processes of climate change already in motion will most likely have their more serious security impacts beyond the next decade, and actions taken within the decade can reduce those longer-term risks.

Implications for Security Analysis

Policy makers can pay attention to only so many warnings. The purpose of this report is to help intelligence analysts determine where to focus and how much attention to pay to the less likely, but potentially significant, developments for security that might result from climate change. This study does not offer recommendations on where or when the U.S. government should act on risks related to climate change. That is a policy choice that will depend on much more than the risks of climate events—or even the risks of humanitarian crises, political instability, violent conflict, or other extreme social or political events that may be influenced by climate change. Rather, the focus of this study is on offering ways to better assess such risks and to anticipate changes in them.

STRUCTURE OF THE REPORT

The next chapter presents the conceptual framework for the project, laying out the key concepts and relationships that provide the structure for analysis. It considers risks as resulting from climate events; exposures of people, places, or important life-supporting systems to these events; vulnerability to these events (susceptibility to harm and the likelihood of effective coping, response, and recovery); and social and political disruptions that may result from responses to these events that are or are perceived to be inadequate. That framework is used in the next three chapters to examine current knowledge about the potential links between climate change and political and social stresses with implications for U.S. national security. Chapter 3 focuses on climate events. It considers what kinds of potentially disruptive climate events can be expected, especially in the coming decade, as a result of climate change. Chapter 4 examines changing exposure and vulnerability to potentially disruptive climate events and takes up the question of what kinds of connections exist between climate events and vulnerability as well as examining processes of coping, response, and recovery after an event. Chapter 5 considers security risks that have been linked to climate change in previous studies and explores the question of what links, if any, exist that connect climate change to political and social stresses and thus to outcomes of obvious security concern, such as violent conflict, pandemics, or disruptive migration. It addresses a continuing discussion in the academic literature about direct versus more complex and contingent relationships. These chapters provide the evidence to support the work in Chapter 6, which takes up a core task of the project: recommending what the intelligence community should be monitoring in order to assess climate–security connections in ways that are useful for policy.

2

Climate Change, Vulnerability, and National Security: A Conceptual Framework

As indicated in Chapter 1, the main focus of this report is a subset of the ways that climate change might affect U.S. national security. In particular, we examine *climate events*, that is, events that are directly connected to properties of climate through deterministic physical or biological mechanisms, and the harm such events cause to human life and well-being when socioeconomic and political systems do not respond adequately. (See Box 2-1 for definitions of key terms used in this report, which are indicated in this chapter by being shown in italics when first introduced.) Climate events, which may be acute or slowly developing, include physical events such as droughts, heat waves, and extreme storms; the ecological consequences of physical changes, such as crop failures and disease outbreaks; and events that may have both physical and biological causes, such as fires in unhealthy forests. Depending on other socioeconomic, political, and environmental conditions and on societal reactions to the disruption caused by climate events, a given climate event may ultimately result in large-scale social and political outcomes that have the potential to affect U.S. national security. This chapter develops several concepts that can help us identify, classify, and discuss the many ways in which climate events may contribute to such social and political outcomes. It also develops a conceptual framework for thinking about the potential security implications of vulnerabilities to climate change, that is, gaps between existing states of adaptation and what the adaptations would have to be to prevent harm from climate events.

> **BOX 2-1**
> **Definitions of Key Terms**
>
> **Adaptation:** the process of adjustment to actual or expected climate and its effects in order to moderate harm or exploit beneficial opportunities (shortened from Intergovernmental Panel on Climate Change, 2012:3). Also, an action that reduces exposure or susceptibility to harm from a potential future event or that increases the likelihood of effective response.
> **Climate Change:** a change in the mean or variability of any of the properties of climate that persists for an extended period, typically decades or longer (adapted from Intergovernmental Panel on Climate Change, 2012:3).
> **Climate Event:** any event that is directly connected to the properties of climate through deterministic physical or biological mechanisms. Such events might be acute (e.g., a storm or heat wave), slowly developing (e.g., a drought or a change in the ecological range of a crop pest), or a combination of the two (e.g., wildfires in a drought-stricken forest). The behavior of climate includes the full range of climate events; the term *weather* normally is applied to short-term climate events.
> **Climate Extreme:** the occurrence of a value of a weather or climate variable above (or below) a threshold value near the upper (or lower) ends of the range of observed values of the variable (Intergovernmental Panel on Climate Change, 2012:3).
> **Coping:** actions taken by individuals and communities using their available resources and typically without the intervention of formal organizations to face and manage adverse conditions, emergencies, or disasters (adapted from United Nations International Strategy for Disaster Reduction, 2009a).
> **Disaster:** a consequence of hazardous physical or biological events interacting with vulnerable social conditions, leading to widespread adverse human, material, economic, or environmental effects that alter the normal functioning of

CONNECTIONS BETWEEN CLIMATE EVENTS AND NATIONAL SECURITY

Figure 2-1 shows our general conceptual framework for thinking about climate–security relationships that involve vulnerabilities to climate events. We developed this framework from an examination of the research literature on the implications of climate change for human well-being and security, and we present it here as a way to think through the connections among various factors and to explore the needs for analysis, monitoring, and projecting potential security threats. We expect that as events proceed, monitoring improves, and understanding increases, the framework will evolve, and it will become possible to formulate it with more precision.

a community or society sufficiently to require immediate emergency response to satisfy critical human needs (Intergovernmental Panel on Climate Change, 2012).

Exposure: the presence of people, livelihoods, environmental services and resources, infrastructure, or economic, social, or cultural assets in places that could be adversely affected (Intergovernmental Panel on Climate Change, 2012:3).

Human life-supporting system: any combination of natural systems and the human activities and institutions that use them to meet critical needs for human life and the well-being of individuals, communities, and societies.

Impact: a consequence for a human or natural system of the interaction of climatic, environmental, and human factors.

Recovery: actions taken by people, communities, and formal organizations in the aftermath of a disruptive event to compensate for the harm caused or to restore altered systems to a more desired state.

Response: the provision of emergency services and assistance, typically by formal organizations such as police, hospitals, and governmental or international organizations during or immediately after a disruptive event in order to save lives, reduce health impacts, ensure public safety, and meet the basic subsistence needs of the people affected (adapted from United Nations International Strategy for Disaster Reduction, 2009a).

Susceptibility: the degree to which a population, community, society, or system suffers or would suffer immediate harm as the result of exposure to a climate event. Thus, susceptibility is an indicator of the extent that an event would create disruptive change in the short term in that population, community, society, or system.

Vulnerability: the propensity or predisposition to be adversely affected (Intergovernmental Panel on Climate Change, 2012:3). Aspects of vulnerability include the *susceptibility* to being harmed and the likelihood of effective *coping* or *response* in the event of harm.

This section briefly describes the framework; its implications are elaborated throughout the report.

Climate Events and Vulnerabilities

People and societies depend for their lives and well-being on a number of complex and interrelated systems that may be affected by climate variability and change. The most important systems are those that meet critical human needs by protecting health and providing water, food, energy, shelter, transportation, and essential commercial products. Each of these *human life-supporting systems* is affected by physical and biological systems, including climate, and by the socioeconomic and political conditions that

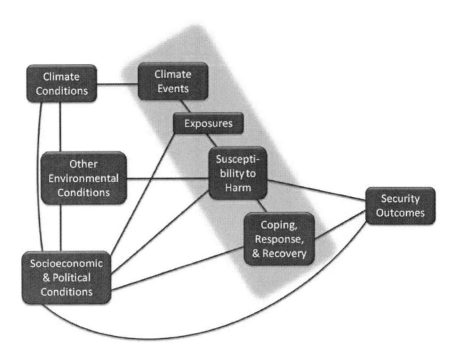

FIGURE 2-1 A schematic model showing the links between climate events and outcomes of national security concern, highlighting the roles of exposure and vulnerability. The shaded area corresponds to the event-exposure-vulnerability model in Intergovernmental Panel on Climate Change (2012). Many of these links involve causation in both directions; however, some of the causal links are much more important than others on time-scales that are important for security analysis.

organize how people and societies interact with those systems to meet their needs. It is important to recognize that some human life-supporting systems, including international disaster assistance, protections against pathogens, and markets for key commodities such as grains and petroleum, are global. This means that climate-related events anywhere that affect these systems have the potential to create disruptions elsewhere on the planet.

People, societies, and their support systems have developed over time in ways that leave them well adapted to a range of variations in physical and biological conditions and events in their environments but vulnerable to harm if events go far enough beyond their usual ranges. Climatic change and climate-related events have the potential to affect each of these life-supporting systems by presenting them with such conditions. There are many ways of providing a general description of how harm might arise. One useful characterization comes from a recent report by the Intergovernmental Panel on Climate Change (Intergovernmental Panel on Climate Change, 2012), which describes harm from climate change as resulting from confluences of climate-related *events*, *exposure* of people or the systems they depend on to those events (i.e., being in harm's way), and the *vulnerability* of those people or things to the events. We refer to this idea as the EEV (for events, exposure, and vulnerability) model of impacts.

In this model the term *climate events* refers to weather and other events that are directly connected to the properties of climate through deterministic physical or biological mechanisms. *Exposure* is defined as the presence of people, livelihoods, environmental services and resources, infrastructure, or economic, social, or cultural assets in places that could be adversely affected. *Vulnerability* is defined in the IPCC report simply as "the predisposition or propensity to be adversely affected" (Intergovernmental Panel on Climate Change, 2012:3).

A large scholarly literature on environmental hazards and climate change elaborates on the concept of vulnerability in ways that we find useful for the purposes of national security analysis (Liverman, 1990; Kasperson and Kasperson, 2001; O'Brien et al., 2004; Wisner et al., 2004; Brooks et al., 2005; Adger, 2006; Eakin and Luers, 2006; Füssel, 2007; Intergovernmental Panel on Climate Change, 2007, 2012; Polsky et al., 2007). In particular, we find it useful to distinguish two elements of vulnerability. One is the *susceptibility* of some aspect of well-being to harm from climate events, taking into account the actions—often called "adaptive actions" in the climate change literature (or "hazard mitigation" in the disaster literature; see, e.g., National Research Council, 2006)—that are taken before an event occurs and that reduce exposure or decrease direct or immediate harm to an exposed system or population compared with what it would have been in the absence of these actions. A system, community, or society is susceptible to harm from a climate event to which it is exposed to the extent that the event creates disruptive change in that system, community, or society.

The IPCC defines *disaster*, an extreme result of events outstripping adaptations, in terms that imply a certain level of susceptibility. Disaster is defined as "severe alterations in the normal functioning of a community or a society due to hazardous physical events interacting with vulnerable social conditions, leading to widespread adverse human, material, economic, or environmental effects that require immediate emergency response to satisfy critical human needs and that may require external support for recovery"

(Intergovernmental Panel on Climate Change, 2012:3).[1] This concept of disaster is very close to our concept of susceptibility: An event is said to produce a disaster when the disruption it causes is severe enough to require an immediate emergency response in order to take care of critical human needs.

The other element of vulnerability that we highlight involves socioeconomic and political reactions to disruptive climate events. Following the terminology common in disaster research (e.g., United Nations International Strategy for Disaster Reduction, 2009a), we distinguish three kinds of reactions: coping, response, and recovery.[2] *Coping* involves actions taken by individuals and communities using their available skills and resources to face and manage adverse conditions, emergencies, or disasters more or less on their own, at least temporarily. Immediately after an unanticipated climate event, coping is the predominant reaction; with some events, it remains the predominant reaction for some time. The term *response* connotes more formal activities and is generally taken to mean "the provision of emergency services and public assistance during or immediately after a disaster in order to save lives, reduce health impacts, ensure public safety, and meet the basic subsistence needs of the people affected" (United Nations International Strategy for Disaster Reduction, 2009a). This definition implicitly recognizes the roles of governmental and other formal organizations, and it specifically refers to short-term activities during or immediately after a disruptive event. *Recovery* involves actions taken by people, communities, and formal organizations in the aftermath of a disruptive event to compensate for the harm caused or to restore altered systems to a more desired state. The difference between response and recovery is not clearly marked, and different areas or groups in an affected area may be in different stages at the same time. Response is particularly important from a security perspective because, as discussed further in Chapter 4, political and social upheavals in the aftermath of environmental disasters typically result more from actual or perceived failures of response by governmental organizations than from the disasters themselves.

---

[1] Climate change can also create hazards when physical events interact with biological systems, such as terrestrial or aquatic ecosystems, in ways that lead to events that affect human well-being. Examples include the spread of pests or pathogens to new places or new hosts, leading to disease outbreaks among humans or food species. Although the drivers of such ecological changes may be physical changes in climate or weather conditions, these events ultimately are hazardous because of their biological aspects. Our definition of disaster includes events that result from connections between physical climate events and their biological consequences.

[2] Because disruptive climate events are typically experienced as natural disasters, the concepts developed over more than half a century of disaster research are particularly useful in our analysis.

Vulnerability can be thought of as the gap between an existing state of adaptation and what it would have to be in order to avoid harm from a potential future event. Alternatively, it can be thought of as the amount of harm to a place, population, or system that would result from a future event, taking into account the available resources for coping, response, and recovery. Vulnerability, therefore, is a function of the event under consideration, the pre-existing conditions of human life-supporting systems, and the likely effectiveness of coping, response, and recovery. It is important to note that disruptive events and reactions to them can significantly alter the vulnerability of an affected place, population, or system. Sometimes, a disaster leaves those affected more vulnerable; however, an ideal recovery process is one that leaves them less vulnerable than before.

Climate Events and Social and Political Disruption

The extent of disruption that occurs as the result of a climate event is determined by the interaction of the event; the preexisting climatic, environmental, socioeconomic, and political conditions; the exposures of places, populations, or life-supporting systems; and their vulnerability, including both susceptibility and deficits in coping, response, and recovery systems. This formulation, which might be called EEV+, expands on the EEV model by including conditions that precede an event and by elaborating the concept of vulnerability to specifically reference the included elements of susceptibility, coping, response, and recovery. These interactions may result in a loss of life and livelihoods as well as in disruptions within any of the major human life-supporting systems, with the possibility of causing harm in places located far from the initiating climate event.

Climate conditions, including averages and variations in climate parameters such as temperature and precipitation, have always generated potentially disruptive events. Some extreme climate conditions have been occurring more frequently in recent decades (Hansen et al., 2012; Intergovernmental Panel on Climate Change, 2012), and the likelihoods of several kinds of extreme climate events are expected to increase—at an increasing rate in most places—in the future (e.g., Intergovernmental Panel on Climate Change, 2012). Climate conditions affect other environmental conditions, including those of hydrological, ecological, and land-cover systems, on both short and long time-scales, and these systems in turn affect climate. Climate conditions similarly interact with socioeconomic and political conditions on long time-scales in the sense that human systems have evolved in response to experiences with climate and that, through the use of various environmental resources, human systems in turn affect climate. Because most of these interactions occur on time-scales of many decades or longer, they are not of major concern for this study.

Of greater concern for this study are the ways in which climate change and climate events interact with the environmental, socioeconomic, and political conditions operating on shorter time-scales that help determine exposure, susceptibility, and the likelihood of effective coping, response, and recovery and thus the potential for social and political consequences that can become national security concerns. Although in the past much of the discussion about climate change and national security has focused on anticipating climate events, in this report we point out that there is also great value in assessing exposure and the elements of vulnerability, which implies monitoring and anticipating the social, political, and economic conditions that determine exposure and vulnerability. It is often the case that the ability to estimate vulnerability is greater than the ability to foresee climate events. The interactions of climate events with other factors are discussed in greater detail in Chapters 4 and 5.

In sum, the potential for climate events to trigger disasters and social or political upheavals depends both on the events and on exposures and vulnerabilities of the places and systems that are affected. To anticipate the damage that climate change may cause and the social disruption that might result, the intelligence community needs not only to anticipate climatic events but also to anticipate exposures, susceptibilities, and the likelihood of effective coping, response, and recovery. Climate science is the main source of knowledge for foreseeing climate events; social science and knowledge of local conditions are the main sources of knowledge for foreseeing exposures, susceptibility, and the likely effectiveness of coping, response, and recovery. We discuss the determinants of future climate events in Chapter 3 and the determinants of exposure, susceptibility, coping, response, and recovery in Chapter 4.

Because the human consequences of climate events are also influenced by pre-existing conditions and by societal factors affecting exposure, susceptibility, coping, response, and recovery, the affected people and countries may not always see the disruption as climate-related. Disputes about the proper attribution of the events can themselves contribute to social disruption. For example, between 2010 and 2012 Pakistan experienced a series of electrical blackouts and shortages of irrigation water, both attributable in part to decreased flows in the Indus River. The decreased flows occurred in the context of a long-term decline in per capita water availability, which by 2010 was less than a third of what it had been in the 1950s (Pakistan Water and Power Development Authority, 2011) as a result of the increasing demands for irrigation water to feed a rapidly growing population, inefficient drainage practices, and possibly inequitable water allocation between regions and uses (Ghumman, 2012a; News International, 2012). Drought arrived on top of these stresses. Protest demonstrations and riots occurred with increasing frequency and intensity during 2010 and 2011, tied mainly to the power blackouts (Ayub, 2011). The blackouts and water shortages themselves were disruptive enough, but, in addition, their cause became a contentious political issue with the potential to inflame Pakistan–India relations. The Pakistani foreign minister blamed the decreased flows on illegal water withdrawals upstream by India; the commissioner of the Indus River System Authority in Pakistan attributed them to climate change (Jafrani, 2010).

Social and Political Disruption and U.S. National Security

Climate events that affect vulnerable places or life-supporting systems can become concerns for U.S. national security in at least two ways. One is by creating social or political stresses or inducing policy actions within or between foreign countries that pose security risks for the United States. Another is by developing into major humanitarian crises that directly create U.S. national security concerns or engage U.S. resources that also have national security missions. In either case, disruptions from climate events in other countries are usually at least one step removed from U.S. national security concerns. The intervening steps may involve violent domestic conflicts or the deterioration of governments' abilities to function in affected countries, cross-border wars, mass international migration, demands on the United States to provide humanitarian relief, or the diversion of U.S. military and other resources from current security missions to disaster response. In short, any effects of climate events and climate change on U.S. national security over the next decade are likely to be indirect, mediated by the susceptibilities of human life-supporting systems to harm in exposed places or through global social or economic processes, the responses of national governments and international institutions to experienced harm, and the social and political consequences of responses that are or are perceived to be inadequate. This means that the potential effects of climate change on U.S. national security cannot be anticipated without considering all of these conditions and potential responses. Security issues may also arise when the response of a government is to suppress popular demands for assistance or to prevent access to aid.

IMPLICATIONS OF THE CONCEPTUAL FRAMEWORK

The framework sketched in Figure 2-1 has several noteworthy implications for security analysis.

"Impacts of Climate Change" Do Not Result Only from Climate Change

The effects of climate change on human well-being and on U.S. national security will depend not only on climatic and environmental conditions, but also on socioeconomic and political ones. Security concerns arise from climate events that prove disruptive. Extreme climate events are sometimes, but not always, disruptive to social and political systems, and climate events that are not extreme can sometimes be very disruptive; the ultimate outcome depends on susceptibilities and reactions to harm. It is important to project the likelihoods of potentially disruptive climate events, but analyses restricted to this, whether globally or at finer scales, will often fail to provide sufficient information to those concerned with human, national, or international security. It is also necessary to consider socioeconomic and political conditions at levels from local to global.

The Picture Changes Over Time

All the elements in the conceptual framework change over time, and they change at various rates. Thus, to anticipate the risks of security concerns arising at any point in the future requires an analysis that examines all of the framework's elements and their interrelationships in the intervening period. Socioeconomic and political conditions often change more rapidly than average climate or many environmental conditions, and this is likely to be the case over the coming decade. When this is the case, the rate of change in the risk of disruption from climate events may be more dependent on changes in the conditions affecting exposure and vulnerability, including actions taken to reduce susceptibility and prepare for disasters, than on changes in climate.

Changing Exposures to Coastal Storms and Floods

A simple example is the growing risk to human populations in coastal areas from storm surge and sea level rise. Climate and environmental change are exposing more land to these hazards, but in many regions rapid population growth and infrastructure development resulting from birth rates exceeding death rates, net migration, and economic development are putting people and property in harm's way faster than climate and environmental change alone. For example, one recent scenario-based analysis estimated increases of population exposure to coastal flooding as a function of climate change and changes in social and economic conditions over the next 60 years in several countries. It concluded that exposure in India would increase from the present level of about 5 million people to about 28 million, with about 13 million of that increase attributed to socioeconomic change and about 10 million to the combination of climate change and land subsidence. In Côte d'Ivoire, Egypt, and Nigeria, almost all the projected increase in population exposure is due to socioeconomic change; in Japan, very little of the projected change in exposure is socioeconomic in origin (Hanson et al., 2011).

Changing Susceptibility to Food Insecurity

In many developing countries economic development and urbanization are making large populations less dependent on subsistence agriculture and local food supplies. This trend will decrease these populations' vulnerability to extreme climate events affecting local crops and meat supplies. At the same time the dependence of low-income populations on imported food supplies provided by global markets may increase their vulnerability to climatic or economic events in other parts of the world that sharply increase the prices of the foods they have come to depend upon. The net effect of these social changes on the well-being of these populations in the face of climate change is likely place-dependent.

Changing Likelihood of Effective Response

Social, economic, and political conditions may also affect the capacity and willingness of governments and societies to respond and mitigate harm to vulnerable populations when disruptive climate events occur. Disaster researchers point out that both "social capital" in the affected communities and formal emergency response institutions and infrastructure play important roles in mediating the net degree of loss, disruption, and stress that result from extreme environmental events, including climate events (see, for example, Aldrich, 2012). Effective response also depends on the economic and other resources available to the governments of the affected populations and on the governments' allocation of those resources. Whether or not climate events become social and political stresses serious enough to destabilize a government or generate violent conflict may depend on whether or not governments' disaster response efforts are perceived to be under-resourced, poorly managed, or characterized by favoritism, corruption, and lack of compassion. (This is discussed in more detail in Chapter 4.) Estimates of risks from climate events may therefore need to take into account the likely future condition of formal and informal response and recovery systems; furthermore, if there appears to be a significant likelihood of inadequate responses, the estimates should consider the likely consequences for social and political systems, including the governments potentially held responsible.

Small Climate Events Could Have Large Social Effects

Thresholds or tipping points have received much attention in the literature of physical climate science. In Chapter 3 we discuss evidence on the likelihood, in the next decade, of crossing important physical thresholds that could lead to a sharply altered climate regime. Less commonly examined are the ways in which changes in human systems might sharply alter vulnerabilities and thus contribute to the potential of even small climate events to have major impacts. Such changes could contribute to social and political stresses, even in an unchanged climate regime, and could have greater effects in the presence of climate change. The following examples illustrate some mechanisms by which this could happen.

Loss of "Slack" in Local Life-Supporting Systems

Relatively slow climatic, ecological, or economic changes can shift the balance of supply and use of natural systems at a local or regional level to the point that adequate supply can be achieved only with favorable climate conditions. The effects may not be noticeable until an unusual climate event occurs, but the responsibility for the impact would in fact lie with the combination of the event and the underlying changes in vulnerability. The decline in water availability in Pakistan in 2010–2012, already mentioned, exemplifies this type of situation. For decades water supplies had been increasingly appropriated to irrigate crops and provide electric power, but this situation did not create a crisis for livelihoods until these slow changes combined with the much decreased water flows in the Indus River to create a situation in which the agricultural and energy systems were highly vulnerable to drought. The Indus water commissioner's claim that the cause of the water shortage was climate change may or may not have been accurate; ordinary climate variation may have been the trigger. Even events within the normal range of climate variability can lead to major disruption if support systems have become vulnerable to them.

Increasing Dependence on Global Markets

Economic development in most countries has generally been marked by a pattern in which livelihoods depend decreasingly on subsistence agriculture and the local manufacture of essential products and increasingly on wage labor and the purchase of necessities in global markets. This transition usually includes a rural–urban shift in national populations as well. Historically, these changes have tended to decrease vulnerability of food supplies to local climate events because when destructive climate events occur locally, necessities can be purchased from places where such events have not occurred. But while direct vulnerability to events that limit local food production has decreased, vulnerability, especially of the lowest-income groups, remains and may be increased with respect to events that limit distribution or that sharply increase prices in global markets for necessities that cannot be acquired locally. Economic globalization thus changes the nature of vulnerability to climate events as well as the degree of that vulnerability (Leichenko and O'Brien, 2008). With globalization, populations become increasingly interconnected via international trade so that it becomes possible, for example, for a climatic event that affects one of the world's grain-producing regions to influence global commodities markets in ways that can seriously affect populations that do not directly experience the climate event. In this way the well-being of households in Lagos or Nairobi can be sharply affected by a drought in Ukraine or the United States. Security analyses should consider the possibility of this sort of phenomenon in commodity markets when assessing the climate vulnerabilities of large low-income populations in key countries.

Climate and Ecological Change

Societal activities are well known to alter ecosystems. Climate change also does this—sometimes slowly, and sometimes as the result of extreme events—with results that may not become evident until an extreme climate event occurs. For example, climate change can alter the ranges of certain species of pests or pathogens, increasing the exposure of human populations or economically important nonhuman species. The expansion of the pine bark beetle in North America is a familiar example. As average temperatures in the region increased, making additional areas suitable for beetle infestations, the beetle expanded its range northward and toward higher elevations (Carroll et al., 2003). The ecological change did not become seriously disruptive to human populations until the increased prevalence of dead trees combined with drought and hot weather to produce major wildfires that affected populated areas. The wildfires in Colorado in 2012, widely described as unprecedented in extent, may have been affected by this process.

Slow climate change could potentially have similar effects on the evolution or distribution of human pathogens (influenza, yellow fever, etc.) or of pests of major crop or livestock species. When one of these pests or pathogens makes contact with a vulnerable population, epidemics, epizootics, or crop failures can spread rapidly, leading to major losses of human life and well-being. Slow processes of ecological change or slow changes in the resistance of host populations to disease organisms could lead to the crossing of a tipping point in vulnerability, at which point the meeting of pest and host populations can set off a highly disruptive chain of events.

Anticipating Disruptive Climate Events: A Trans-Disciplinary Problem

The climate sciences help in anticipating climate events. Various other scientific disciplines are engaged in understanding some of the many societal processes that affect exposure and vulnerability and can therefore help in estimating the disruptive capacity of future climate events. These disciplines need to be more fully engaged with the climate sciences in order to assess the disruptive potential of possible climate events. For example, demography has been particularly successful at forecasting exposures by estimating future human populations from fertility and mortality data and trends and from migration data and trends. Researchers in the development and planning fields can provide useful estimates of infrastructure development that can help with estimating the exposure of property to climate events known to be prevalent in particular areas. Forecasts of economic growth and economic well-being of populations can be useful for anticipating the degree to which extreme events will create serious disruption or suffering. Engineering analyses can estimate the ability of physical infrastructure to withstand possible extreme events. Ecological analyses of habitat change for pests and pathogens can help in foreseeing outbreaks of some human, animal, and plant diseases. There are also social scientific bases for estimating capacity and willingness to cope and respond in environmental emergencies, although some of these, such as for forecasting social capital, are at early stages of development. In addition, there is a body of research literature on the effects of the quality of disaster response on the stability of governments. We discuss these issues further in Chapter 4.

STRATEGIES FOR SECURITY ANALYSIS

As the discussion so far makes clear, there are many plausible scenarios by which climate change, climate events, and their interactions with non-climate environmental conditions and socioeconomic and political changes might set processes in motion that create national security concerns for the United States. It is also clear that the likelihood that any specific scenario will arise is highly uncertain. This section considers strategies that the U.S. intelligence community might use to better inform national security decision makers with regard to potential risks related to climate change.

The problem might be phrased simply as one of determining which potential futures are important enough to worry about. Policy makers have limited cognitive bandwidth, so they can pay attention to only so many warnings. After some important event has occurred it is often easy in retrospect to identify the relevant precursors. Before the event, however, it is often much more difficult to separate the signal from the noise.

We reemphasize the point made in Chapter 1 that the appropriate standard of evidence for considering security risks is different from the standards of evidence in fundamental science. Intelligence and security actions are often warranted to reduce the risks from events whose likelihoods are low or cannot be predicted with confidence because the phenomena are too complex or poorly understood or because of the importance of human agency in shaping the course of events. Many of the risks associated with climate change have these characteristics. Nonetheless, the relevant sciences sometimes can develop useful estimates of changes in the likelihoods of certain kinds of events and consequences. Security analysis needs to apply a risk-based analytic approach to recommending action, such as the one briefly described in Box 2-2.

There are several general approaches for implementing a risk-based climate–security analysis. One takes a *forecasting* approach: Analysts project the likelihoods of disruptive events and bring the high-risk forecasts to policy makers' attention. Applied to climate change, this approach would need to involve the forecasting of climate events as well of as societal conditions that alter exposure and susceptibility to harm from those events and that affect the ways that governments, societies, and other social institutions respond when climate events create social disruptions. Forecasting requires an understanding of the key variables that need to be included in the forecast as well as a theoretical framework that specifies how current conditions and trends are linked to future outcomes of concern.

This approach potentially has the advantage of bringing to decision makers' attention a range of scenarios worth worrying about. In particular, because risk is the product of likelihood and impact, this approach would likely bring both high-likelihood/medium-impact events and low-likelihood/high-impact events to policy makers' attention. Both types of events may be relevant for assessing the impact of climate change on social and political stresses of security interest. An important drawback of this approach is that the ranking of risks—and thus the selection of the events that fall above and below the threshold for policy makers' attention—may be highly sensitive to erroneous estimates of both likelihood and consequence in ways that are poorly understood or even unnoticed a priori. Such errors may occur in the forecasting of every relevant factor—in the forecasting of climate events, of exposure and susceptibility, and of reactions to disruptive events. The difficulties of forecasting all of these things given the current state of knowledge has led us to put a low priority on using this approach.

A second approach emphasizes *early warning*. Analysis can suggest early warning indicators of significant events in countries of interest. Analysts might, for instance, identify early indicators of climate conditions that are likely to lead to serious economic or social consequences for the affected populations or indicators of a lack of political or social capacity to cope with or respond to such consequences. They might also develop

> **BOX 2-2**
> **A Risk-Analytic Approach to Climate Events and Stresses**
>
> Risk is typically defined as the severity of an undesired outcome multiplied by the likelihood of its occurrence. Climate change alters both the likelihood of occurrence and the likely severity of certain events that may degrade human life-supporting systems. Changes in these systems may in turn alter the likelihood and severity of social disruption, stress on political systems, and events of potential importance to U.S. national security—violent internal or international conflict, state failure, and so forth.
>
> Social conditions and social changes affect the exposure and vulnerability of human populations and societies to climate events and thus the risks associated with these events. They also shape the responses of social systems when such events occur, further influencing the risks of political and social stress.
>
> It is essential to think about climate and security in terms of risks because it is beyond the capacity of today's science to predict specific climatic events or their social or political consequences years or even months in advance. However, the natural and social sciences can help the intelligence and national security communities understand whether, where, and to what extent the risks of these effects are changing and how to reduce them.
>
> The security risks posed by climate change are multidimensional. The overall risk may depend on attributes of:
>
> *Climate events:*
> 1. Types of climate events (e.g., floods, crop failures, and disease outbreaks)
> 2. The likelihoods of the events

indicators of exposure, susceptibility, or the likelihood of successful coping and response, either of socioeconomic systems or of the resource bases on which human populations in critical regions depend.

A third approach emphasizes the *analysis of system vulnerabilities*. Such an analysis might be focused, for example, on the social and political capabilities and weaknesses of particular regimes both generally and in the face of expected climate events. It would ask how capable a particular regime is of dealing with its routine challenges and then consider how able it would be to deal with additional stresses, specifically those that might arise from future climate events. The focus for a particular country, region, or system should be on the examination of potentially disruptive climate events that have a reasonable likelihood of arising there in the coming years—that is, a set of plausible worst cases—along with various ways that the regime might address such problems (e.g., hardening infrastructure,

3. The magnitude and extent of the events
4. The places and times that the events occur
5. Whether events arrive singly or in clusters

Exposures to the events:
6. The populations, communities, and infrastructure affected
7. The life-supporting systems affected (e.g., food, water, energy, health)
8. The social and political systems affected

Susceptibilities of exposed systems to harm from events:
9. The extent to which these systems in the affected communities and countries will suffer harm upon exposure

The likelihood of effective coping, response, and recovery:
10. The capacity of exposed people and communities to cope
11. The capacity and willingness of various social institutions to respond and support recovery of the affected systems

Security implications of ineffective reactions to disruption:
12. The reactions of the affected populations and social institutions to the adequacy or inadequacy of response
13. The ability of governments to cope with post-event reactions, and consequences for well-being on larger social and political systems

Because risks are so varied in kind and in their implications, it is useful to think in terms of a risk profile—the shape of the risk on all the above dimensions—as a basis for systematic analysis.

putting in place general purpose rapid response capabilities, or suppressing demonstrations by unhappy citizens). A vulnerability approach might also be applied to systems within countries, e.g., of the water supply in Pakistan or the food supply in China. The "stress testing" that we recommend in Chapter 6 mainly follows this approach.

A fourth approach emphasizes *policy vulnerability analysis*. Analysts would consider a specific existing or proposed policy, future states of the world in which that policy would fail to meet its goals, and the policy responses that would ameliorate these vulnerabilities. They would then organize scenarios to illuminate for policy makers the trade-offs among responses to maintain the viability of the policy. For instance, such an analysis might note that climate changes that go beyond certain thresholds would likely be sufficiently disruptive to the functioning of the government in Pakistan to significantly undermine the ability of current U.S. policy to achieve its policy goals there; then identify policy changes that might reduce those vulnerabilities (e.g., by raising the threshold at which climate change would harm the ability of the United States to achieve its goals); and, finally, provide information to decision makers that would help them decide whether or not to adopt those policy changes. This approach has the advantage of generally being less likely to misestimate risks than the forecasting approach, but it requires analysts to analyze the implications of specific policy options.

In practice, a mix of approaches will likely prove most useful, because each approach has the potential to miss important pieces of the security situation. For example, a forecasting approach might identify stresses that might arise in a country of interest and create general conditions of social disruption, but it might not carefully consider the implications for specific U.S. policy goals in the country. A policy vulnerability analysis will need help from the forecasting or early warning approach to estimate the thresholds at which current policies might lose viability.

All of these analytical approaches can be better informed by the monitoring of a wide range of variables, including those describing climatic and other environmental factors as well as socioeconomic variables; the monitoring priorities may in some cases depend on the analytical approach being used. Forecasting can require the monitoring of a large number of all these types of variables, both to support forecasts and to build and validate the theories on which risk estimates are made. Early warning obviously requires the monitoring of climatic variables, but it also depends on the monitoring of other environmental and socioeconomic variables that could make climate events disruptive in countries of interest. Analysis of system vulnerabilities requires the monitoring of regime capacities and capabilities, but it also requires monitoring the various climatic and socioeconomic factors that might make climate events stressful for the system. Similarly, policy vulnerability analysis requires the monitoring of U.S. policy goals and priorities, but it also requires monitoring the range of conditions that could make policies vulnerable. In Chapter 6 we offer recommendations for monitoring that can support these analytic strategies.

3

Potentially Disruptive Climate Events

Earth's climate provides the environment in which humanity has evolved and in which human societies have expanded and thrived. It also periodically generates events that disrupt those societies—in some historic cases, apparently causing the failure of entire civilizations, although in many of those cases considerable dispute exists about the precise cause (Butzer, 2012). Climate events are disruptive when they harm people and the environmental, social, and economic assets that people depend on and when societies are unable to cope, respond, and recover effectively. Disruption thus depends on the conjunction of events with human vulnerability. However, major climate events are not always seriously disruptive. Past civilizations have sometimes faced very serious stress from environmental change without collapsing (Butzer, 2012; Butzer and Endfield, 2012).

When climate is changing, potentially disruptive climate events occur at different rates, with different intensities, and perhaps in different places from what people would expect from past experience. Those responsible for anticipating the risks, including the intelligence and security communities, turn to climate science in the hope of getting useful estimates of changing overall risks of future events of interest and forecasts of when and where such events will occur on time-scales longer than those of normal weather forecasts.

This chapter focuses on climate events, a key element in the conceptual framework presented in Chapter 2 for considering the connections between climate change and outcomes of security concern. It briefly summarizes the scientific bases for providing risk estimates and forecasts for potentially disruptive climate events on time-scales up to a decade or so. We consider the potential for major, abrupt climate change as well as for single climate events and clusters and sequences of such events that could be sufficiently disruptive as to raise concerns about U.S. national security. The next two chapters examine how such events might disrupt social, political, and economic systems sufficiently to raise national security concerns.

THE SCIENCE OF CLIMATE PROJECTION

The predictability of weather and climate varies with the time-scale; moreover, predictive efforts face different challenges at different time-scales. Weather forecasting focuses on predictions at time-scales up to about two weeks and is based on the premise that the atmosphere behaves according to a set of deterministic equations such that if the initial state of the atmosphere is known, its evolution can always be determined. Changes or errors in the initial state limit predictability on longer time horizons; two weeks is normally considered the limit of atmospheric predictability.

Climate models are based on the same basic set of equations that predict shorter-term weather variations, but they also include terms that represent a coupling of the atmosphere with the ocean and land surfaces, which inherently have memories of climate longer than the atmosphere does. Climate models have a coarser resolution than weather prediction models, which limits the level of accuracy in their simulations of atmospheric and ocean dynamics and their interactions with climate. Instead of forecasting actual day-to-day changes in weather over a period of a week or more, climate models concentrate on simulating the processes that govern the interannual and longer-term climate variability of the coupled ocean–atmosphere–land system.

To date, climate prediction has focused mostly on two time-scales: seasonal and centennial. Seasonal predictions, like weather forecasts, are dependent on initial values, and thus their ability to make predictions relies on information provided by initial ocean conditions (Latif et al., 2010), particularly sea surface temperatures, which strongly influence atmospheric circulation. The greatest contributor to predictive skill on a seasonal time-scale has been an understanding of the dynamics of the El Niño–Southern Oscillation (ENSO), which influences the yearly variability of rainfall and temperatures over broad sectors of the globe and even global mean temperatures.

On centennial time-scales, the evolution of climate remains chaotic and irregular and depends on external changes in radiative forcing (the influence of a factor, such as solar radiation or anthropogenic changes in atmospheric composition and land cover, on the balance of incoming and outgoing energy in the Earth-atmosphere system). Thus on such time-scales, climate projections are sensitive to assumptions about how future radiative forcing will change as a function of energy consumption, land cover, and other drivers of change in Earth's radiation balance. On such time-scales, the system loses its memory of the initial conditions, and the future trajectory of the coupled climate system is strongly determined by the external forcing.

Recently observations and models have suggested that it might be possible to make decadal climate predictions by using knowledge about regularities in the natural climate system on that time-scale, especially conditions involving the state of the ocean. However, climate projections based on emission scenarios indicate that decadal-scale variability is also influenced by the accumulated impacts of anthropogenic radiative forcing. Decadal-scale predictions therefore require information both about radiative forcing (e.g., levels of greenhouse gases and aerosols) and about the current observed state of the atmosphere, oceans, cryosphere, and land surface (Hurrell et al., 2010).

Understanding and making predictions of decadal variability is still very much in its infancy, and evaluating these predictions is a major challenge. It is straightforward to verify daily weather forecasts through statistical and historical data, but verification is more difficult for decadal forecasts. Observational records are simply not consistent enough or long enough to quantify prediction skill. Even the basic characteristics and mechanisms that describe climate on a decadal scale are poorly documented and not well understood. Testing models against observed climate variability provides some means of verification and thus can offer some confidence in using these models to simulate future climate, but even those efforts are hindered by a lack of subsurface ocean observations and satellite data.

As part of the work for the next Intergovernmental Panel on Climate Change (IPCC) report, modeling centers have coordinated decadal hindcast and prediction experiments covering the period from 1960 to 2035 in what is known as the Coupled Model Intercomparison Project Phase 5 (CMIP5). Part of the work involves an atmosphere–ocean general circulation model with a resolution of approximately 50 kilometers (31 miles) being run to make decadal predictions out to 2035. Lower-resolution versions of the same model will be run with coupled carbon cycles and biogeochemical processes to help quantify the magnitude of important feedbacks that will determine the degree of climate change in the second half of the 21st century (Hurrell et al., 2009).

A better knowledge of the initial conditions needed to initialize the models properly should increase the skill of predictions of the climate system on varying time-scales. An increased understanding of the physical mechanisms that govern climate variability is also critical. In sum, although physical climate science is able to project climate trends based on scenarios for increases in greenhouse gases (GHGs) and to estimate changes in the likelihoods of occurrence for some kinds of climate events in the coming decade, the ability to anticipate specific climate events on that time-scale is still in its infancy. Improving our understanding of the ENSO cycle and its broader climatic implications currently offers the best route for extending our ability to make skillful forecasts of the risks of climate events from the current limit of a few weeks into the future to a few months or even one or two years. It will also be important to improve understanding of how the GHG-forced changes in climate might change the characteristics of climate variability, such as ENSO.

Despite their inability to predict specific events for the coming decade, climate models can still provide useful information about the range of plausible climate outcomes that could result from the combination of external forcing and variability of the climate system due to internal processes over the next decade. For example, several recent studies have evaluated the distribution of 10-year trends within model simulations that were forced with increasing concentrations of greenhouse gases (Easterling and Wehner, 2009; Santer et al., 2011). The purpose of these studies was to address claims that the lack of a significant positive global temperature trend over the 10-year period following 1998 refutes projections from climate models. And, indeed, the models did show that even under scenarios with relatively high greenhouse gas emissions, decades with a zero or negative temperature trend occur in more than 10 percent of the individual 10-year segments within models.

Figure 3-1 displays the probability distribution functions of 10-year trends from Easterling and Wehner (2009) for global mean temperatures, and it extends their method to look at the probability distributions of 10-year trends for a number of regions for the future (see Appendix C for an explanation of the method used for the regional projections). These functions were calculated using model simulations from the CMIP5 database under a scenario for the first half of the 21st century that assumes a "business as usual" rate of greenhouse gas increase of about 1 percent per year (known as the RCP8.5 scenario). The functions represent differences between projected average annual surface temperature, globally and for various regions, for the coming decade and the average temperature at the start of the decade, expressed in degrees Celsius. The estimates apply to any decadal start date up to 2040. Much as Easterling and Wehner (2009) found for global temperatures, each region has a reasonable chance of cooling over a period of a decade because of natural variability that can act to counteract the warming trend. At the same time, the upper tails of these distributions show that the increase in average global temperature could be as high as 1.8°F (1°C) in a single decade and the increase in average regional temperatures could be as high as 3.6°F (2°C) or more in a single decade—for example, about a 40 percent chance of an increase of at least 1.8°F (1°C) and about a 10 percent chance of an increase of at least 3.6°F (2°C)

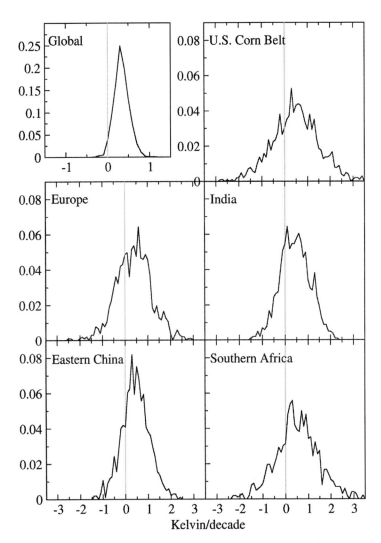

FIGURE 3-1 Distributions of probabilities for average annual surface temperature, globally and for the regions shown, for the coming decade, relative to the average temperature at the start of the decade, expressed in degrees Celsius. Although the most likely 10-year trend in each region is for warming (that is, most of the area under each curve is to the right of the zero degree line), each region has a reasonable chance of overall cooling for the coming decade. Each region also has a reasonable chance of warming by more than 1.8°F (1°C) and some chance of warming more than 3.6°F (2°C) in the coming decade.
SOURCE: Committee analyses.

in the U.S. corn belt. Although the scenarios of strong and rapid warming over a decade have low probability, the potential implications for society are large enough to deserve much more consideration by climate and security analysts. To our knowledge, similar work has not yet been published for plausible decadal trends at regional scales or for climate variables other than surface temperature.

Although the foregoing analysis suggests very useful predictive skill for changing likelihoods of some kinds of events, even projections that have relatively low predictive skill can be useful to the security community. Low predictive skill is especially likely to be characteristic of projections of climate parameters with low signal-to-noise ratios (e.g., with high interannual variability) and for very low-frequency events, including the most extreme events. Such events deserve special consideration by security analysts when and where the fundamental science suggests that they will increase in likelihood and the data are not yet adequate to rule out this possibility. In such situations, it may be wise for security analysts to consider what-if scenarios involving increased frequencies of occurrence for such events.

ABRUPT CLIMATE CHANGE

Climate models currently used to project climate change over the 21st century generally indicate that there will be a gradual response to increased greenhouse gases and other climate forcing, suggesting quite a low likelihood that major changes will occur within the coming decade. Nevertheless, the simulations do show that abrupt changes are possible in various regions where a particular variable crosses a threshold (for example, if precipitation lowers sufficiently to cause a transition from forest to grassland, or grassland to desert, or if temperature increases enough to cause a transition from snow- or ice-covered surfaces to snow- or ice-free ones). The observed climate record, particularly the paleoclimate record, shows that abrupt or step changes in climate do occur, even within a time span of a decade or less. The possibility of rapid step changes in climate is of great concern in terms of social and political stresses because, with less time to adjust, it is usually harder for societies to cope and respond effectively.

Abrupt climate change is generally defined as occurring when some part of the climate system passes a threshold or tipping point resulting in a rapid change that produces a new state lasting decades or longer (Alley et al., 2003). In this case "rapid" refers to timelines of a few years to decades. Abrupt climate change can occur on a regional, continental, hemispheric, or even global basis. Even a gradual forcing of a system with naturally occurring and chaotic variability can cause some part of the system to cross a threshold, triggering an abrupt change. Therefore, it is likely that gradual or monotonic forcings increase the probability of an abrupt change occurring.

Many cases of rapid change in climate have been observed in the instrumental and paleoclimatic records. The instrumental record is short—only about 130 years in length—and as a consequence does not adequately sample the full spectrum of climate variability. However, even with this limited sampling some examples of rapid regional change can be found in the observed record. For example, strong warming (greater than 4°C) occurred in parts of the Arctic during the 1920s, with most of the warming occurring in the late 1920s (Alley et al., 2003). The Dust Bowl years of the 1930s in the central United States provide another example of a rapid shift, in this case with the reversal also occurring in less than a decade. On a much larger geographic scale, the climatic effects of the abrupt shift in 1976–1977, mainly in sea surface temperatures in the Pacific Ocean, have been documented in a number of global and hemispheric analyses (Alley et al., 2003).

The paleoclimatic record indicates that numerous rapid changes occurred in the period after the last deglaciation (i.e., the past 10,000 years). Many of these changes were larger than those recorded in the instrumental record. Examples include abrupt shifts to large-scale drought that were similar to, but more intense and longer lasting than, what was experienced during the 1930s Dust Bowl years. Tree ring records of drought in the western United States show abrupt, long-lasting mega-droughts that came on rapidly and lasted for a decade or more, covering most of the region. It has been argued that large-scale mega-droughts were responsible, at least in part, for the decline and collapse of the Mayan civilization in Mexico (Hodell et al., 1995), the Tang Dynasty in China (Yancheva et al., 2007), and the Anasazi civilization in the southwest United States (U.S. Climate Change Science Program and Subcommittee on Global Change Research, 2008). Such conclusions are hard to support conclusively, however. For example, Butzer and Endfield (2012) judge the Mayan case to be controversial.

What can cause rapid climate change? Some rapid changes are initiated by a cooling process, others by a warming process. Some rapid changes are temporary and reversible, others cause one-way shifts that persist for decades or longer. One recurring cause is massive releases of sulfate aerosols into the atmosphere, which can result in substantial regional or global cooling. Historically the main cause of such releases has been volcanic eruptions. Large volcanic eruptions, such as the 1815 Tambora eruption that led to the world-wide cooling event called the "year without a summer" in 1816, occur on average about once every 1,000 years. Super-eruptions occur on average about once every 10,000 years, with the last two having been in Toba in Indonesia 74,000 years ago and Oruanui in New Zealand 26,000 years ago. It is thought that the Toba super-eruption led to a global "volcanic winter" lasting for several years that was approximately 7.6°F (4°C) colder than today and up to 27°F (15°C) colder in the high latitudes.

Some theories suggest that the Toba eruption and ensuing volcanic winter wiped out most of our human ancestors, leaving only a small population of perhaps a few thousand (Self, 2006).

The above examples of rapid climate change do not fully meet the definition of abrupt climate change given by Alley et al. (2003) because after each of these rapid changes, the climate returned to its previous state. However, it is possible for a monotonic change in certain aspects of the Earth system to cross a tipping point so that climate change not only would be rapid but also would shift the system to a different baseline. For example, as the area covered by Arctic summer sea ice decreases, exposing more ocean surface, the result is an overall darkening and loss of reflectivity for Earth's surface in that region, which leads to an increase in the absorption of solar radiation; this in turn warms the ocean waters, which tends to increase melting, creating a positive feedback loop. Thus the result of a loss of summer sea ice could be a lasting change in the energy balance of the planet. However, because there are feedbacks that lead ice extent to recover, some analysts argue that a lasting change in the energy balance from this mechanism is unlikely in the coming century (Tietsche et al., 2011).

Could significant and lasting abrupt climate change occur, either globally or regionally, in the next decade or so? The possible scenarios most frequently mentioned include the loss of Arctic summer sea ice, GHG release from melting permafrost, the melting of Greenland glaciers, the breakup of the West Antarctic ice sheet, and a change in the strength of the Indian summer monsoon (e.g., Lenton et al., 2008). At a smaller spatial scale, the loss of specific mountain glaciers or glacier systems, the drying of lakes, and shifts in biomes are likely types of abrupt changes for particular regions.

The likelihood of any of these events occurring by the mid-2020s is difficult to estimate for a number of reasons, including the fact that global climate models do not simulate abrupt climate shifts well, which makes it hard to project when a change in some part of the climate system that might induce a broader abrupt change (e.g., regional or hemispheric temperature or precipitation) will reach a critical value. Lenton et al. (2008) provide estimates of critical values for a number of "tipping elements," which are defined as features of the Earth system large enough (at least subcontinental in scale) to have pronounced impacts on the climate system if an abrupt change occurs in that feature. Examples would include the Greenland ice sheet and the Amazon rain forest. Even for those tipping elements where critical values are defined (e.g., for rapid melting of the Greenland ice sheet, approximately a 5.4°F [3°C] increase in the local air temperature), estimating a probability that the critical value will be reached in the next couple of decades is difficult if not impossible. Most climate scientists would probably judge the likelihood of an abrupt change that would have large and long-lasting destabilizing impacts on climate and human society by the mid-2020s to be quite small. However, because of gaps in knowledge about the mechanisms and tipping points for the processes involved, it is difficult to assign a level of confidence to judgments about the preconditions for, or timing of, an instability affecting, for instance, a major section of a large ice sheet. Moreover, there may be other processes in the Earth system, not yet identified, that have tipping points that could lead to abrupt climate change. Because of such gaps in knowledge, the possibility of such events occurring in the next decade or so cannot be totally discounted.

Although the likelihood of such abrupt climate change scenarios happening in the next decade or so appears to be quite small, it cannot be estimated accurately. The consequences of some abrupt climate changes, if they occurred, could be quite severe. For example, droughts are a recurring event in the climate system, with a major drought occurring in some part of the world at least yearly. Some drought conditions, such as those that have continued off and on for the past decade in the western United States, could be harbingers of a mega-drought. If that proved to be the case in the western United States, the potential impacts on water supplies as well on as ecosystems, both natural and managed, in that region would be enormous. To the extent that the possible tipping elements leading to such major changes are known, it would be important to monitor those elements and the factors that affect them. It would also be important to monitor changes in the social, economic, and political factors that affect the size of the exposed populations, their susceptibility to harm, the ability of the populations to cope, and the ability of their governments to respond. Where potentially affected areas are important producers of key global commodities such as food grains, it would also be important to assess the effects of climate-induced supply reductions on global markets and vulnerable populations.

SINGLE EXTREME EVENTS

The fundamental science of climate change suggests that continued global warming will increase the frequency or intensity (or both) of a great variety of events that could disrupt societies, including heat waves, extreme precipitation events, floods, droughts, sea level rise, wildfires, and the spread of infectious disease. Underpinning many of these extreme events is an acceleration of the global hydrological cycle. For each 1.8°F (1°C) increase in the global mean surface temperature, there is a corresponding 7 percent increase in atmospheric water vapor. Because warm air holds more water vapor than cool air, this leads to more intense precipitation. Essentially, warm air increases evaporation from the ocean and dries out the land surface, providing more moisture to the atmosphere that will rain out downwind. Water vapor is also a powerful naturally occurring greenhouse gas. As such it is the source of a very strong positive feedback to the coupled climate system that amplifies any external forcing by a factor of approximately 1.6.

This section discusses trends in some extreme climate events over the past half century, science-based expectations of the futures of these types of events, and the prospects for using the science of climate change to estimate the changing likelihood of such events and to forecast their occurrence.

Trends in Extreme Climate Events

The frequency of extreme high-temperature events driven by global warming is increasing faster than would be the case if only the mean temperature were increasing because the variance of the temperature distribution is increasing as well (see Figure 1-1). Extreme weather and climate events have been responsible for a rapidly increasing loss of lives, well-being, and economic assets in recent decades because of the confluence of the events themselves with increases in the numbers of people and value of property exposed and vulnerable to the events. It has been difficult to determine conclusively whether damaging climate events themselves have yet been increasing in frequency or intensity enough to be detected in trends of damage, normalized for nonclimate factors. We discuss this issue further in Chapter 4.

Climate Change and Extreme Climate Events in the Coming Decade

Effects of Climate Change on Extreme Events

The frequency and intensity of extreme events are particularly hard to project because, among other things, there are by definition few of them, which makes it hard to validate predictive models against experience (see Appendix D for more detailed discussion of statistical issues and methods for assessing the probabilities of occurrence of extreme events). Analyses have, however, converged on a number of expectations for this century, as noted in a recent National Research Council review (2010a). In this century, it stated, "the frequency and intensity of heat waves is projected to continue to increase, both in the United States and around the world," and "the frequency of cold extremes and the number of frost days will decline in the middle and high latitudes" (p. 223). It is also projected that "the fraction of rainfall falling in the form of heavy precipitation events will increase in many regions" (p. 224). Furthermore, "[r]ecent model projections indicate growing certainty that climate change could lead to increases in the strength of hurricanes, but how their overall frequency of occurrence might change is still an active area of research," and "projections indicate that the area affected by drought will probably increase in the decades ahead and that the number of dry days annually will also increase" (p. 265). The models are not highly specific about the timing of these changes, however.

A subsequent IPCC special report on extreme events (Intergovernmental Panel on Climate Change, 2012) summarized available evidence on observed and projected extremes of a variety of important types of climate events globally and at the regional level. Table 3-1 summarizes observed trends in several of these.[1] The level of confidence in an observed trend neither implies nor excludes the possibility of changes in an extreme at other geographic scales. The report also offers projections of trends in these events, along with statements about the levels of confidence that the scientific community places in the projections.

Considering the difficulty of validating projections of extreme climate events, it would be a mistake to conclude from the lack of confidence in a projection of change for any type of extreme event that one can prudently act as if there will be no change. When there is good fundamental science behind an expectation of change—for example, in the frequency of extreme high-temperature and high-precipitation events or the likelihood of droughts—combined with noisy data or a small number of events for model validation, there may be sufficient reason for the intelligence community to develop and consider the security implications of scenarios in which the extreme event parameter changes in the direction suggested by the fundamental science.

Effects of Predictable Climate Variation on Extreme Events

Earth's climate includes various regular cycles that may make it possible to anticipate an increased likelihood of climate events of concern months or longer in advance. Other than the succession of the seasons, the best understood of these is the ENSO (see Box 4-1). ENSO affects weather on most of Earth's surface in cycles that last two to seven years. The largest and most predictable impacts are in the tropics. The warm phase of ENSO "is usually accompanied by drought in southeastern Asia, India, Australia, southeastern Africa, Amazonia, and northeast Brazil, with fewer than normal tropical cyclones around Australia and in the North Atlantic. Wetter than normal conditions during El Niño episodes are observed along the west

---

[1] The confidence levels used in Table 3-1 are based on three scales: evidence and agreement; confidence; and likelihood (Mastrandrea et al., 2010). The summary terms used to describe the available evidence were limited, medium, or robust, and the degree of agreement was described as low, medium, or high. Levels of confidence were very low, low, medium, high, or very high. Likelihood (probability) was described as virtually certain (99–100%), very likely (90–100%), likely (66–100%), about as likely as not (33–66%), unlikely (0–33%), very unlikely (0–10%), and exceptionally unlikely (0–1%).

TABLE 3-1 Observed Changes in Extreme Weather and Climate Events Since 1950

Event	Global Change	Confidence	Regional Changes	Confidence
Cold days and nights	Decrease	Very likely	Decrease at the continental scale in North America, Europe, and Australia	Likely
Warm days and nights	Increase	Very likely	Increase at the continental scale in North America, Europe, and Australia	Likely
			Warming trend in daily temperature extremes in Asia	Medium
			Warming trend in daily temperature extremes in Africa and South America	Low to medium, due to insufficient evidence
Length or number of heat waves			Increase in many regions with sufficient data	Medium
Heavy precipitation events			More regions have experienced increases than decreases, although there are strong regional and subregional variations	Likely
Tropical cyclone activity (intensity, frequency, duration)	Unclear			Low confidence after accounting for past changes in observing capabilities
Drought			Intense and longer droughts in southern Europe, west Africa	Medium
Extreme coastal high water	Increase related to an increase in mean sea level			Likely, at global scale

SOURCE: Adapted from Tables 3-1 and 3-2 in Intergovernmental Panel on Climate Change (2012).

BOX 3-1
The El Niño–Southern Oscillation Phenomenon

The El Niño–Southern Oscillation (ENSO) phenomenon is a multi-year cycle marked by changes in relative atmospheric pressures at sea level across the tropical Pacific Ocean and changes in the strength of the Pacific trade winds and the temperature of the ocean's surface in the central and eastern Pacific. This mode of climate fluctuation has been linked globally to devastating droughts, extreme rainfall events, and the suppression of hurricanes in the Atlantic Ocean, among other interannual climatic changes, and therefore has a significant impact on global society. Recent studies have also linked ENSO to the natural variability of the carbon cycle (Jones et al., 2001). An El Niño event (the warm phase of the cycle, as indicated by sea surface temperature) is characterized by a weakening of the Pacific trade winds, a warming of the ocean's surface in the central and eastern Pacific, and smaller differences in tropical sea level pressures between the eastern and western tropical Pacific Ocean. Opposite conditions occur in the cold phase, or La Niña. The warm events often last 12 to 18 months and historically occur every 2 to 7 years, although recent anthropogenic climate changes may possibly contribute to more frequent and intense El Niño events. The term ENSO is used to describe the full range of coupled ocean–atmosphere climate variability in the tropical Pacific Ocean.

Over the course of many decades of research, great strides have been made in understanding, monitoring, and predicting ENSO and its effects on climate. During an El Niño event, the warmest waters of the ocean, usually west of the international dateline, spread eastward into the eastern tropical Pacific, a distance one-third the circumference of the planet, and significantly affect the global atmospheric circulation. Via a process known as teleconnection, ENSO is often linked to global extreme precipitation events, either drought or flooding, in many far-flung places. ENSO also appears to play a large role in yearly and multi-year variations in sea level rise, while larger ENSO events can affect decadal trends in sea level by shifting the distribution of rainfall between land and the ocean. ENSO also affects marine and terrestrial ecosystems. During a warm ENSO phase the primary productivity in the eastern tropical Pacific is severely reduced because of weaker upwelling off the coast of Peru. The reduction in productivity affects the population and location of marine mammals, sea birds, and commercial fishing. Year-to-year variability in global atmospheric carbon concentrations is dominated by the ENSO cycle (Rayner et al., 1999). During El Niño, equatorial upwelling decreases in the eastern and central Pacific, significantly reducing the supply of carbon dioxide to the surface (Feely et al., 2006). As a result, the global increase in atmospheric carbon dioxide noticeably slows down during the early stages of an El Niño. Coastal regions are particularly affected by ENSO variability.

coast of tropical South America, subtropical latitudes of western North America, and southeastern America" (Intergovernmental Panel on Climate Change, 2012:155). The cold phase, known as La Niña, generally shows opposite anomalies. The effects of ENSO can also be felt in the other ocean basins and on all continents (Dai et al., 1998). Recent research has shown that different phases of ENSO are associated with different frequencies of short-term weather events such as heavy rainfall and extreme temperatures in the affected regions (Intergovernmental Panel on Climate Change, 2012).

These regularities in the effects of ENSO on climate events allow for skillful short-term climate predictions on time-scales from a season to a year by coupling atmospheric general circulation models with ocean general circulation models initialized with observations of the state of atmosphere and ocean. Forecast skill decreases away from the equator, however. This seasonal climate prediction has now become operational at many of the world's major weather prediction centers. (NOAA, for example, offers seasonal outlooks online at http://www.cpc.ncep.noaa.gov/products/predictions/90day/ [accessed November 13, 2012].) Prediction on time-scales from years to decades is still very much in the research realm, so it is not yet possible to estimate the eventual skill that might be achieved in forecasting at these time-scales.

The fact that ENSO drives unusual weather patterns on several continents at the same time demonstrates that extreme weather patterns in different regions may not be independent in a statistical sense, suggesting the possibility that extreme climate events may cluster in time, a topic discussed more fully in the next section.

The question of whether and how anthropogenic climate change may be altering the ENSO cycle is being actively examined by climate scientists (Intergovernmental Panel on Climate Change, 2012). Dai et al. (1998) noted a shift in ENSO events in the mid-1970s toward more warm events, which coincided with record high global temperatures and drought anomalies that were greater than expected. There is some evidence linking these changes in ENSO events to intensified droughts in some drier regions since the 1970s, while the extent of wet areas has declined during this period. These changes are qualitatively consistent with the observed increases in greenhouse gases in the atmosphere, which act to enhance the hydrological cycle. The IPCC special report on extreme events (Intergovernmental Panel on Climate Change, 2012) describes systematic changes in ENSO behavior that have been observed over the past 50 to 100 years, but it concludes that it is not clear what role, if any, increased greenhouse gases have played in this phenomenon. All models used in that assessment predict continued ENSO interannual variability in the future no matter what the change in average background conditions. However, the changes in ENSO interannual variability differ from model to model based on subtle changes in the physical parameterization schemes of the models (Collins, 2000). The models are not consistent in their projections of ENSO amplitude or frequency in the 21st century, and it is not yet possible to tell which models' results are most credible. The IPCC study concluded that "it is not possible at this time to confidently predict whether ENSO activity will be enhanced or damped due to anthropogenic climate change" (Intergovernmental Panel on Climate Change, 2012:157). Climate scientists continue to explore links between increased greenhouse gas concentrations and ENSO because the fundamental principles of atmospheric science indicate that greenhouse warming will drive changes in ENSO.

It seems reasonable to infer from this research that ENSO is likely to change in coming decades, even though we cannot at this point determine which parameters will change or to what extent. In our judgment it would be worthwhile for the intelligence community to consider the security implications of a few of the scientifically plausible scenarios for how ENSO might change because changes in the amplitude and nature of ENSO constitute one of the few components of the coupled climate system capable of having a global synchronized impact. However ENSO changes, there will be simultaneous consequences in many places. For example, if the future will bring stronger El Niño events, as some models project, the results will likely include drought in Australia and excess precipitation across parts of the southern tier of the United States occurring in the same year. Such a confluence of events could affect global grain production in ways that could be disruptive, depending on other conditions affecting food supply and demand.

Conclusion

Over the next decade the most likely scenario for single extreme events is a continuation of the trends of recent decades—probably a slow rate of change for now, but with the possibility of a faster one later. Many of these trends involve the continuation of trends that are already being observed, such as those summarized in Table 3-1, such as a warming of days and nights and a trend toward heavier precipitation, with wet areas tending to get wetter and dry areas drier, and with more of the precipitation concentrated in heavy events, particularly in certain regions. The effects of the ENSO cycle, superimposed on these longer-term climate trends, will exacerbate some extremes and dampen others in complex ways. In addition to these events, there remains the possibility of unprecedented extreme events that might occur as a result of abrupt climate change or other climatic phenomena discussed later in the chapter.

CLUSTERS OF EXTREME EVENTS

By a *cluster of extreme events* we mean several extreme climate events appearing close in time but not necessarily in space. Clusters of extreme events are a concern from a national security perspective because U.S. government resources and those of other international actors deployed to deal with a security or humanitarian concern related to the first event in a cluster might be unavailable or less available to deal with a second or subsequent extreme event.

Clusters of extreme events may occur as a result of a random co-occurrence of extreme events in different places that have different causes. They may also result from large-scale climate processes that serve as common causes of events in disparate places that seem superficially to be unrelated. An example of such an event cluster was the conjunction of a heat wave and drought in Russia and floods in Pakistan in 2010. In July and August of that year, western Russia suffered from a major drought and heat wave that resulted in a major loss of life and crops as well as large forest fires. Simultaneously Pakistan was experiencing major flooding that also resulted in a major loss of life and property. These two events were linked by more than just their proximity in time. The meteorological pattern that led to the Russian heat wave, in which the large-scale upper-level wind flow developed a strong and persistent ridge, also contributed to the development of the meteorological pattern that resulted in the Pakistani floods—a downstream, leading trough (Lau and Kim, 2012). The fact that these two extreme events corresponded in time with each other and with a single larger meteorological pattern was unusual but not totally unexpected. Circulation events like this one, which cause some event clusters, are known to occur but are not well resolved in current climate models.

A key question for security analysis is whether such event clusters are non-random, indicating a teleconnection phenomenon in which the events in the cluster are intrinsically linked, making their joint probability of occurrence greater than the individual probabilities multiplied together. If so, when one of these events occurs, the likelihood of the other is increased. Knowledge of systematic connections of this type could be valuable for security analysis because it could aid in the development of security risk scenarios.

Generally this sort of clustering of events can occur when the large-scale atmospheric flow around the mid-latitudes of the northern and southern hemispheres, which contains a series of about four to six meanders, or waves (called Rossby, or planetary, waves), develops specific kinds of stable patterns, resulting in the simultaneous occurrences of drought, drought and flood, or flooding events in different parts of the globe. Herweijer and Seager (2008) show that the simultaneous occurrence of drought in North America, parts of Europe, South America, and Australia has resulted from the development of Rossby wave patterns, likely a response to specific patterns of cold sea surface temperatures in the tropical eastern Pacific Ocean (La Niña). Much as with the recent Russian drought/Pakistani flooding cluster, when drought has occurred over North America and Europe (including western Russia), unusually wet conditions have occurred over Pakistan and northern India as well as over other parts of the globe (Herweijer and Seager, 2008). The ENSO phenomenon creates systematic connections of climate conditions that affect large parts of the planet more or less simultaneously, leading to predictable clusters of extreme climate events during strong El Niño and La Niña phases (see Box 4-1). Such phenomena suggest that the simultaneous occurrence of drought in parts of the northern and southern hemispheres coincident with flooding in other regions is a distinct possibility in the next decade or so. Other clusters of extreme weather or climate conditions may also become more likely than in the past.

Related to the concept of event clusters is the notion of compound events, which occur when different kinds of climate events are linked in the same place. An obvious example is the conjunction of drought with wildfires or crop failures. The first event drives the second, but they are different in that different people or groups are susceptible to damage from different parts of the compound event and different organizations may be involved in response. In Chapter 5 we discuss an example of the conjunction of drought, heat wave, and forest fire in Australia in 2010.

If climate events and extremes were independent in a statistical sense, the likelihood of a cluster or a compound event of any size could easily be estimated mathematically. But as the above example makes clear, extreme events in different parts of the world can be driven by common underlying forces and thus have an intrinsic relationship such that when one such event occurs, the likelihood increases that other extreme events linked to them by common causes will also occur. In statistical language, such events are called dependent.

We conducted two exploratory analyses in which we used data from 1901-2002 to examine statistical relationships between pairs of conditions that climate science indicates may be dependent on underlying large-scale climate patterns (see Appendix D for details). In one case, the correlation between heat in Russia and heavy precipitation in Pakistan, we found no general association between the two phenomena in the century-long record, suggesting that the association in 2010 was without recent precedent. In the other, an association between droughts in regions of the southwestern United States and a region in Argentina and Uruguay, we estimated that the probability of drought exceeding a 10-year return level in one of the regions given a similarly serious drought in the other was almost three times what it would be if the events were statistically independent and that for more serious droughts, the dependence effect was even stronger.

These results are preliminary, and conclusions should be tempered by questions about the accuracy of the data, the limited number of data points (102 annual observations), and the small number of previous examples that were extreme in both variables. The results suggest, however, that event clusters of this sort could be increasing in likelihood more rapidly than the underlying single events and that by combining climatological and statistical analysis, it may be possible to develop better estimates of the likelihoods of occurrence of extreme event clusters of interest. However, the scientific analysis on this matter is still in its infancy.

SEQUENCES OF EVENTS

A climate condition or event can sometimes precipitate or facilitate a series of other physical and biological events, each linked to the others through deterministic processes, that together promote conditions of potential security concern. We can illustrate the concept of climate event sequences, although not the potential national security implications, with a sequence of events that took place recently in western North America.

The recent pine bark beetle outbreaks in western North America have their roots in multiple mechanisms, but climate change is believed to be a factor driving at least some of them (Bentz, 2008). Elevated temperatures, particularly consecutive warm years and elevated minimum temperatures, can speed up reproductive cycles and reduce cold-induced beetle mortality, both of which increase beetle populations. Moreover, shifts in precipitation patterns and associated drought can promote bark beetle outbreaks by weakening trees and making them more susceptible to beetle attacks. Droughts, combined with forest die-off from beetle outbreaks, in turn increase the amount of fuel in forests, increasing the susceptibility of forests to fire, especially during heat waves, which are likely to be longer and more intense with climate change. The ongoing climate change, because of its warming, also increases evaporation and thus intensifies the risk of fire (see Figure 3-2). Dead forests provide an opportunity for fires to spread more widely. The extreme forest fires in Colorado in the summer of 2012 occurred in one of the highest-risk regions in the western United States. Severely burned forest lands are also more prone to erosion in storms (e.g., Benavides-Solorio and MacDonald, 2005), indicating that forest fires increase the risks of soil degradation and of mudslides. Climate change may thus be playing at least four different roles in this dynamic: It promotes bark beetle infestations, weakens trees, dries the environment, and creates weather conditions conducive to fire outbreak. These conditions, connected in sequence, increase the risks of major forest fires and their hydrological and human consequences.

The kinds of linkages involved in this example and in other plausible

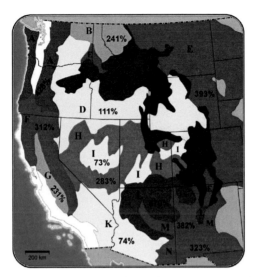

FIGURE 3-2 Map of increased risk of fire in the western United States as a result of rising temperatures and increased evaporation. The figure shows the percentage increase in burned areas in the West for a 1.8°F (1°C) increase in global average temperatures relative to the median area burned during 1950–2003. For example, fire damage in the northern Rocky Mountain forests, marked by region B, is expected to more than double annually for each 1.8°F (1°C) increase in global average temperatures. With the same temperature increase, fire damage in the Colorado Rockies (region J) is expected to be more than seven times what it was in the second half of the 20th century.
SOURCE: National Research Council (2011a).

event sequences are in general poorly understood and therefore not very predictable. Nevertheless, the example suggests that there may be a very large number of plausible ways by which climate change could set in motion a sequence of events in ecological, hydrological, or other deterministic physical and biological systems that could become seriously disruptive and that might create security concerns.

GLOBAL SYSTEM SHOCKS

Climate events occurring in one part of the world have the potential to affect other parts of the world through important, globally integrated systems other than climate itself. One example is the potential influence of climate events on the world supply—and therefore the prices—of international traded commodities, such as grains. By this mechanism an event such as the 2012 drought in the central United States, still developing as this is being written, could affect world corn or wheat prices in ways that make essential foods unaffordable for populations in Africa or Asia. Another example of a global system shock would be constraints on the availability of humanitarian aid for a country because aid providers are responding to situations elsewhere in the world. Yet another would be a climate event that altered the distribution of a major pathogen affecting people or staple crops. These examples, which are discussed in greater detail in Chapter 4, indicate that there are numerous ways in which climate events could create shocks to integrated global social, economic, health, or technological systems and thus have effects far removed geographically from where the events occur.

SURPRISES ARISING FROM POORLY RESOLVED CLIMATE DYNAMICS

Many extreme events, such as hurricanes, tornadoes, other severe storms, floods, heat waves, and wild fires, occur on the time-scale of days to about two weeks. A key question for climate research is how such extreme events will change within a warming climate. With such warming the statistics of weather are no longer stationary, and the linkage between weather and climate emerges as a research priority. Today's climate change models are generally considered to provide an adequate representation of the large-scale secular trend in climate. However, gaining a better insight into how climate change will affect extreme weather requires that high-resolution numerical weather prediction models (and their inherent fast physics, such as cloud–radiation–precipitation interactions) be run in climate mode. This presents a major computational and resource challenge and has been the reason for the call for a seamless approach to weather and climate forecasting (World Climate Research Programme, 2009). Such a unified approach was first implemented by the U.K. Met Office in 1993. In the United States the National Centers for Environmental Prediction of the National Oceanic and Atmospheric Administration developed the Coupled Forecast System (CFS) in 2004 as a lower-resolution climate version of its Global Forecast System, a weather forecast system for short-term seasonal climate prediction. However, this system has yet to be run on decadal to centennial time-scales in response to GHG forcing. Hence the behavior of extreme events in a non-stationary climate cannot be fully described and projected until climate change models are run with the spatial resolution and physical processes of numerical weather prediction models. Until that time, extreme climate surprises should be expected to be more the rule than the exception.

CONCLUSIONS AND RECOMMENDATIONS

Physical climate science has developed some skill at estimating the changing likelihoods of the occurrence of certain kinds of climate events, such as heat waves and certain precipitation anomalies, at a decadal timescale and at a global and, in some cases, a continental or subcontinental geographic scale. However, the ability to foresee specific climate events on a decadal time-scale at the level of medium-sized countries is still in its infancy. Predictive skill at such time horizons and levels of resolution depends on having a more extensive observational system than currently exists and on developing an improved understanding of interannual and decadal processes of climate variability that can be incorporated into predictive models. Estimating the risks of extreme climate events—and especially estimating the places they will occur—is particularly challenging because predictive skill is harder to acquire and to validate for infrequent events and because the frequency and character of such events may change as climate trends continue.

The current state of understanding does not suggest that the distributions of single climate event types in particular places will change sharply in the coming decade, but neither can it preclude such a possibility with high confidence. It is safe to say that extreme climate events will change in their frequencies, intensities, and probably also their locations in the coming decade. The most likely scenario is a continuation of the temperature and precipitation trends of recent decades, probably with a slow rate of change for now, but with the possibility of a faster one later. However, the effects of the ENSO cycle superimposed on longer-term climate trends will exacerbate some extremes and dampen others in complex ways. In addition, there remains the possibility of unprecedented extreme events or conjunctions of events that might occur as a result of abrupt climate change or other climatic phenomena (for example, a more rapid rise in sea level if the melting of ice sheets were to accelerate).

Conclusion 3.1: *Given the available scientific knowledge of the climate system, it is prudent for security analysts to expect climate surprises in the coming decade, including unexpected and potentially disruptive single events as well as conjunctions of events occurring simultaneously or in sequence, and for them to become progressively more serious and more frequent thereafter, most likely at an accelerating rate. The climate surprises may affect particular regions or globally integrated systems, such as grain markets, that provide for human well-being.*

Some events or conjunctions of events may arise from connections within the Earth system, such as aspects of the ENSO cycle and other continental-level phenomena that are only beginning to be appreciated and understood. Climate change has moved the physical Earth system into conditions that are unprecedented in recorded history, and we do not understand the complexities of the relationships among the elements of the system well enough to make skillful forecasts of specific events. It is therefore prudent to expect that the likelihoods of such events occurring will increase in the coming decade in most places and that the rate of change will continue subsequently to increase. Models are an increasingly powerful tool for examining the likelihood of changes in means, extremes, and variability of climate.

It makes sense for the intelligence community to apply a scenario approach in thinking about potentially disruptive events that are expectable but not truly predictable. For example, available climate models sometimes disagree about the direction of a climate trend even when the fundamental science strongly suggests that change is likely. In such situations it may make sense to consider the security implications of two or more plausible trends as a way to anticipate risks. It will also be valuable for the intelligence community to have improved forecasting ability for subcontinental climate events and for event clusters and sequences. An improved monitoring of factors that might provide early warning of potentially disruptive events would also be valuable. We discuss monitoring issues in Chapter 6.

Recommendation 3.1: *The intelligence community should participate in a whole-of-government effort to inform choices about adapting to and reducing vulnerability to climate change. It should, along with appropriate federal science agencies, support research to improve the ability to quantify the likelihoods of potentially disruptive climate events, that is, single extreme climate events, event clusters, and event sequences. A special focus should be on quantifying risks of events and event clusters that could disrupt vital supply chains, such as for food grains or fuels, and thus contribute to global system shocks.*

This research should include efforts by climate scientists to improve fundamental understanding of the effects of climate change on the likelihoods of extreme climate events and also efforts to apply the methods of extreme value statistics to these problems, particularly the problem of estimating the likelihoods of clusters of extreme climate events that are dependent on the same underlying climatic processes. Having improved likelihood estimates for single and clustered extreme climate events would help in defining climate event scenarios for countries, regions, and systems that could be used as the basis for the climate stress tests that we discuss in Chapter 6.

4

How Climate Events Can Lead to Social and Political Stresses

This chapter examines a number of factors identified in the conceptual framework in Chapter 2 as affecting whether and how climate events can lead to significant disruptions in societies or political systems. Understanding these factors—exposure and susceptibility to harm from hazards plus coping, response, and recovery after disruptive events occur—has long been a central concern of several communities of scholars and policy makers. For example, researchers and practitioners concerned with natural disasters have developed concepts, measurement methods, and data to improve understanding of disaster preparedness and response and of the social and economic consequences of floods, wildfires, storms, and various other climate-related events. Researchers and practitioners in such areas as public health, water management, famine and food insecurity, and humanitarian relief have also developed knowledge that is relevant for understanding how climate events may lead to significant social and political consequences. Knowledge from these fields is critical for understanding how and under what conditions climate events can disrupt societies. It has also been central to progress made in understanding vulnerability and adaptation to climate change in the work of the Intergovernmental Panel on Climate Change and the U.S. Global Change Research Program. In this chapter, we briefly review knowledge and insights from these fields that are relevant to security analysis and also review the implications for assessing the potential of climate change to influence social and political stresses.

LOCAL AND DISTANT EFFECTS

We begin by noting that climate events can be socially disruptive both where they occur and elsewhere. The local disruptions are familiar: Storms, floods, heat waves, droughts, and the like have their most obvious consequences where they occur. However, as we noted in Chapter 2, climate change can result in events that create shocks to globalized systems that support human life and well-being and that can therefore affect populations far from where the climate events occur. Here we discuss the susceptibilities of several key globalized systems to harm from such shocks.

Global Food Systems

Under normal conditions the globalization of markets, access to humanitarian relief, and public health systems all tend to reduce the susceptibility of countries and their populations to local climate risks. For example, one of the first responses of governments to expected shortfalls in domestic production is to secure food imports (Timmer, 2010). Yet these global institutions have evolved in, and in some sense been calibrated to, a climate regime that may differ in important ways from the climate of the coming decades. For example, a key feature of commodity markets is the maintenance of stocks that buffer the impact of short-term fluctuations in supply or demand. The levels of these stocks are determined by several factors, including storage costs, interest rates, and the perishability of the commodity, but a key factor is the expected volatility of supply (Wright, 2011). If climate change were to increase the chance of relatively large shortfalls in global production, stocks based on historical expectations of supply variability could be insufficient. Similarly, the capacity of countries to provide humanitarian or public health assistance is related to historical experience with the level of aid needed around the world.

Relatively little peer-reviewed literature has focused on how climate changes in the coming decades and the ability or inability of institutions to adapt to these changes will affect the likelihood of global systemic shocks to food systems, such as rapid price increases. There is some basis for expecting that indicators that aggregate over large areas, such as global food production or global incidence of humanitarian disasters, will be more affected by global climate trends than outcomes in any single location simply because the "signal" of climate change relative to natural variability tends to be clearer at larger spatial scales (Intergovernmental Panel on Climate Change, 2007). For example, Figure 1-1 in Chapter 1 shows that the total fraction of land area experiencing extremely warm temperatures (more than three standard deviations above average) has risen sharply in the past three decades, even if some individual regions have not seen dramatic warming.

A shortfall in food supply that arose from multiple bad harvests around the world and that was large by historical standards would not necessarily result in rapid price increases, given that other important factors affect price fluctuations. If a sufficient number of preceding good harvests had helped to build up stocks, if growth in biofuel demand related to energy policies slowed or reversed, or if a global recession reduced aggregate food demand, supply shortfalls could have relatively little influence on global markets. However, when bad harvests occur in an already tight market, this will generally result in large increases in food prices, as analyses of recent episodes of high prices in 2007–2008 and 2010–2011 have emphasized (Abbott et al., 2008, 2011; Wright, 2011). Policy responses to the initial shortfalls, such as export bans designed to stabilize domestic markets, then often act to further amplify price changes.

In light of recent food price increases, there has been a renewed interest in the effects of high international food prices on domestic prices and social and political stresses. One clear finding is that domestic prices in many countries change substantially less than global prices, partly because of exchange rate variability and partly because of policies aimed at stabilizing domestic prices, such as tariff adjustments, export restrictions, and the use of government storage (Dawe, 2008; Naylor and Falcon, 2010). Nonetheless, in 2008 and 2011 most countries witnessed significant increases in prices versus historical levels, with consequences for local producers and consumers.

The susceptibility of national populations to global price increases depends in large part on the countries' net trade positions: Major importers will generally be hurt, and exporters will benefit. The MENA (Middle East and North Africa) region is the main area of the world that relies on food imports for a large (more than 30 percent) fraction of calories consumed. Wheat prices are especially important in the MENA, given that nearly half of all calories consumed in some countries are from wheat (Food and Agriculture Organization of the United Nations, 2012). A recent World Bank study (Ianchovichina et al., 2012) found that MENA countries are highly vulnerable to global food price shocks. Sub-Saharan Africa is also relatively dependent on food imports, with roughly 40 percent of rice and 70 percent of wheat consumption derived from imports (Naylor and Falcon, 2010).

Because the prices of basic commodities such as bread or flour are often subsidized, demonstrations and even riots frequently occur in response to efforts by governments to reduce subsidies, for example as part of structural adjustment policies. In general these disturbances are contained without an impact on the regime, even if there may be significant violence or property damage. The issue with regard to climate change is whether that pattern could change and that the countries most vulnerable to food price increases could become vulnerable to severe social and political unrest. Unfortunately, there is very little in the peer-reviewed literature concerning the links between food price increases and political unrest. One notable exception is a recent working paper (Bellemare, 2012) that presented an econometric analysis of global data since 1990 and found that high food prices were significantly correlated with political unrest related to food prices, with the latter measured by counting the number of news stories with at least five mentions of terms related to food and riots (or their synonyms).

Interest in the topic has increased in recent years, particularly within the community concerned with food security, spurred on by the question of whether rising food prices played a role in sparking the unrest of the "Arab Spring" of 2011. It is worth noting that the rapid food price increases in the MENA during this period were not driven by local weather conditions, but by events around the world including a severe heat wave in Russia. A report by Lagi et al. (2011) notes that clusters of unrest in the MENA region in 2008 and early 2011 both began immediately after the United Nations Food and Agriculture Organization food price index passed a value of 210. Although they do not identify a causal link between high food prices and riots, the authors argue that a food price index value of 210 represents a simple potential predictor of increased unrest in food-importing countries. Breisinger et al. (2011) find that the unrest was preceded by a drop in food security across the MENA, and Ciezadlo (2011) emphasizes the role that food subsidies have played in popular attitudes toward regimes throughout the region. Johnstone and Mazo (2011) draw connections between climate events (which reduced global food production in the years preceding 2011) and the uprisings, describing climate change as a potential "threat multiplier" in the case of already unstable situations. All of these analyses are careful to note that drawing direct causal links between food prices and political instability is not possible, but they argue that food prices must be considered along with political and cultural factors in explanations of the uprisings.

Global Energy Markets

Like the food system, markets for energy commodities have become increasingly integrated globally over recent decades. In the case of petroleum, this integration is essentially complete: There is one global market that determines prices of crude petroleum, linking producers and consumers around the world (Yergin, 2006). Integration is also increasing, although lagging significantly behind petroleum, for other energy commodities such as electricity (Jamasb and Pollitt, 2005; Boëthius, 2012), natural gas (Siliverstovs et al., 2005), and coal (Wårell, 2006). Thus, possibilities for energy system shocks to have global impacts in the coming decade lie primarily in the petroleum sector.

The integration of petroleum markets was stimulated by desires to safeguard the supply of oil from manipulation by political actors in the wake of Organization of Petroleum Exporting Countries embargoes in the 1970s (Yergin, 2006). A consequence of this integration was that by the 2000s the petroleum system had become so complex and interconnected that, as one study concluded, "a disruption in one part of the infrastructure can easily cause severe discontinuities elsewhere in the system" (International Institute for Strategic Studies, 2011:21). Furthermore, the sensitivity of the system has increased because of a rapid growth in global petroleum consumption that has not been matched by a corresponding increase in production. The result has been an extremely tight market, with petroleum supplies not significantly greater than demand (Gupta, 2008). This "demand shock" (Yergin, 2006), led by the emerging economies in China and India, has left global markets volatile and very sensitive to disruptions in supply (Patrick, 2007; Gupta, 2008; International Institute for Strategic Studies, 2011).

In this tight, sensitive market, climate events that disrupt the production or distribution of oil could lead to price spikes across the global energy market. Several types of climate events could cause such disruptions. Tropical storms and the increased storm surges that result from sea level rise and, in some cases, land subsidence, can disrupt production, refining, and transport of petroleum. For example, one-third of U.S. petroleum refining and processing facilities are located in coastal areas vulnerable to storms and flooding (Schaeffer et al., 2012). Similar infrastructure vulnerabilities exist in Europe and China as well (International Institute for Strategic Studies, 2011). In addition, because offshore oil and gas platforms are generally not designed to accommodate a permanent rise in mean sea level, climate-related sea level rise would disrupt production (Burkett, 2011). The effects of Hurricanes Katrina and Rita in 2005 illustrate this potential. The storms disrupted oil and gas production from offshore rigs, refining at facilities in the coastal zone, and transportation via port facilities and pipelines, causing a spike in global prices (U.S. Department of Energy, 2005; Yergin, 2006; Schaeffer et al., 2012). The pattern repeated, although with a smaller magnitude, when Hurricanes Gustav and Ike hit the Gulf Coast region in 2008, destroying drilling rigs and disrupting refineries (Paskal, 2010).

Other climate events could also affect the global oil market. Oil refining requires large amounts of water for cooling purposes; hence, reduced water availability during a drought would reduce refining capacity. If drought is accompanied by increased temperatures, refineries will require more cooling water to operate, potentially exacerbating the situation (Schaeffer et al., 2012). Also, Arctic energy infrastructure (pipelines and drilling operations) is vulnerable to damage from subsidence caused by melting permafrost (Paskal, 2010; International Institute for Strategic Studies, 2011).

Climate change thus entails some increased likelihood of petroleum supply disruptions, and such disruptions are likely to affect market prices, potentially causing price shocks. The likely magnitude and duration of these price shocks, however, has not been addressed in the research literature. There has been some analysis of their potential macroeconomic effects. Hamilton (2003, 2008), reviewing six decades of oil price and macroeconomic data, reported a very strong relationship between oil price shocks and recessions. To the extent that economic disruptions drive political instability (see, e.g., Alesina et al., 1996), it is plausible that an oil price shock could increase instability, particularly in a situation that is already politically sensitive. However, little research to date has directly addressed the political impacts of energy price shocks, whether caused by climate-related supply disruptions or other factors. These possibilities deserve more careful empirical analysis, particularly as energy markets continue to tighten with increased consumption from Asian nations and as risks increase of climate events disrupting energy supplies.

Strategic Product Supply Chains

Over the past few decades the globalization of many industries has been accompanied by a streamlining of their supply chains in order to reduce costs. However, as a 2012 World Economic Forum publication noted, "the focus on cost optimization has highlighted the tension between cost elimination and network robustness—with the removal of traditional buffers such as safety stock and excess capacity" (p. 10). Climate events can thus be a source of major disruptions in world markets for critical non-food commodities. Such events are counted as one of the major risks to be addressed in the U.S. *National Strategy for Global Supply Chain Security*, released in January 2012 (White House, 2012).

Although not attributed to the effects of climate change (Peterson et al., 2011), the floods in Thailand in 2010–2011 illustrate how an extreme climate event that stresses a government's ability to respond can have global consequences. Much of Thailand, including portions of the capital Bangkok and its surrounding manufacturing districts, was flooded for extended periods between July 2011 and January 2012. The flooding resulted in more than 800 deaths, affected 13.6 million people, damaged 7,700 square miles of farmland, and caused more than $45 billion in economic losses (World Bank, 2011).

Thailand is a flood-prone country with an extensive system of dams, drainage canals, levees, and other flood-control systems, but a series of events in 2011 overwhelmed this system. The most immediate event was the abnormally high rainfall that year. In March 2011, for example, the rainfall in northern Thailand was more than three times the mean level. The abnormally low rainfall of 2010 was another contributing factor. The initial response of dam managers to the intense rainfall of early 2011 was to store the water in the depleted reservoirs, building capacity and preventing early flooding. But when major rains unexpectedly continued, the reservoirs filled, and the dams had to release the water, resulting in flows too large for the downstream drainage canals and levees. This overflow was exacerbated by the many decades of deforestation that have taken place in northern Thailand, which allowed a greatly increased runoff from the rains and helped to overwhelm the downstream flood defenses.

The Thai government suffered significant criticism for what many saw as its mismanagement of the situation. The government was criticized for its forecasts that underestimated the scale of the flooding, for its management of the upstream dams that exacerbated downstream flooding, and for poor communications. Once the flood waters began to overwhelm the existing flood defenses, the government launched many emergency responses, including evacuations, the placing of sandbags, and the diversion of water from some channels to others. In one case the government placed hundreds of anchored boats in one river so that their propellers could help push water toward a second river. As the damages increased, many of these responses were criticized as inadequate. In addition, resistance appeared in some localities where flooding had increased due to barriers designed to protect neighboring communities. Some people ripped down the sandbags that they saw as unfairly diverting flood waters to their areas.

The floods also caused significant disruption to regional and global supply chains. Manufacturing parks located near Bangkok supply parts for the worldwide automobile and electronics industries. One-third of the world's hard drives and high percentages of other key computer components are built there (Connor, 2012). Many of these Thai manufacturing areas were covered by up to 3 meters of water, causing parts shortages worldwide. Even the computer firms located elsewhere in Thailand that escaped the flooding found they could not get critical parts. Production is not expected to fully recover until 2013 (Mearian, 2011). In the meantime, component prices rose as suppliers attempted to stockpile what was available and manufacturers found they could not get the parts they needed. The flooding of automotive parts production facilities forced Honda and Toyota to slow production lines in many countries (Fuller, 2011).

Other Global System Effects

Climate events might also put stress on global health systems in various ways, most of them hard to predict. As discussed in the next chapter, climate change is expected to alter the ranges of disease vectors or pathogens in ways that expose large human populations to diseases to which they have not been previously exposed. This could lead to a rapidly increasing demand for treatments and supplies that may not have been adequately stockpiled. If such health problems arise in combination with a disruption of supply chains for critical inoculations or medications, the potential for a severe health crisis could grow dramatically. Again, the effects might be felt far from the locations where the climate events occur. Climate events, especially when they occur in clusters, can also stress the capacity of international disaster response and humanitarian relief systems and thus cause harm in places that are not directly affected by the events but that need international assistance for other reasons.

Such shocks to integrated global social, economic, health, or technological systems are likely to have different effects in different places. It is reasonable to expect that they would be most disruptive in countries that are dependent on imports of the products of the global system that is shocked and in places or among populations that are particularly susceptible to harm if the availability of the outputs of those systems is restricted by price or policy.

EXPOSURES

Since the 1980s the number of recorded natural disasters related to weather and climate events has roughly doubled, while the number of those related to geophysical events, such as earthquakes, tsunamis, and volcanic eruptions, has neither increased nor decreased (Munich Re, 2012). Reported losses from global weather- and climate-related disasters also increased over the past few decades, mainly because of monetized direct damages to assets, with the amounts of losses varying greatly from year to year and region to region (Intergovernmental Panel on Climate Change, 2012). Since 1980 annual disaster losses have ranged from a few billion dollars to more than $200 billion (in 2010 U.S. dollars), with the greatest losses coming in 2005, the year of Hurricane Katrina. Loss estimates are lower bounds because many impacts, including the loss of human lives, cultural heritage, and ecosystem services, are difficult to monetize and so are poorly reflected in these estimates. Middle-income countries with rapidly expanding asset bases are particularly vulnerable to changes in the frequency, intensity, geographic range, and duration of extreme events. From 2001 to 2006 disaster losses were about 1 percent of gross domestic product (GDP) for middle-income countries, 0.3 percent of GDP for low-income countries, and less than 0.1 percent of GDP for high-income countries (Intergovernmental Panel on Climate Change, 2012). Most fatalities from extreme weather and climate events (95 percent) occur in developing countries.

The major causes of the long-term increase in economic losses from weather- and climate-related disasters have been the increasing exposure of people and the increased value of economic assets in exposed regions. Calculations on the long-term trends in economic disaster losses adjusted for wealth and population increases, which have been conducted in an effort to separate the effects of change in the frequency or intensity of damaging climate events from the effects of increased exposure and vulnerability, have not attributed the increase in losses to climate change alone, but neither has a role for climate change been excluded (Neumeyer and Barthel, 2011; Intergovernmental Panel on Climate Change, 2012). These studies have not accounted well for vulnerability or for adaptation efforts, and they are limited by poor data availability. Settlement patterns, urbanization, and changes in socioeconomic conditions have influenced the observed trends in the exposure and magnitude of harm from climate events (Intergovernmental Panel on Climate Change, 2012). In particular, rapid urbanization and the growth of megacities, especially in low-income countries, have led to the emergence and growth of highly exposed and highly susceptible urban communities.

As discussed in Chapter 3, projections for the next few decades indicate that there will likely be a continuation of current trends, with greater changes in the frequency, intensity, duration, and spatial extent of some extreme events by the end of the century (Intergovernmental Panel on Climate Change, 2012). Who and what is exposed to an extreme weather or climate event depends on the event. For example, many regions are susceptible to flooding following heavy precipitation events, although the flooding and resulting damage can take a number of forms. The people and places most susceptible to harm when there are changes in the frequency, intensity, duration, and spatial extent of extreme events depend on the event and on local factors. For example, a typhoon coming ashore in the Philippines has very different consequences from one of similar strength striking Japan (United Nations International Strategy for Disaster Reduction, 2009b).

It is important to consider the possibility of compound events (see Chapter 3), such as what occurred in South Australia in January 2009 (Murray et al., 2012). An unprecedented heat wave occurred during a multi-year drought, exposing the area to some of the highest temperatures on record. In central Victoria the 12-year rainfall totals were approximately 10 to 20 percent below the 1961–1990 average (Australian Government, 2009). In Victoria, during the week of the heat wave there was a 25 percent increase in total emergency ambulance dispatches and a 46 percent increase over the three hottest days. There were 980 deaths during the four days of the heat wave, compared with an average of 606 per year over the previous five years. A few days after the heat wave, temperatures spiked again, and the forest fire danger index reached unprecedented levels. High winds caused a power line to break, sparking a wildfire that became one of the largest, deadliest, and most intense firestorms in Australia's history; 173 people died. The bushfires also destroyed almost 1,660 square miles of forests, crops, and pasture as well as 61 businesses. The Victorian Bushfires Royal Commission conservatively estimated the cost of the fire at AUS$4.4 billion (Parliament of Victoria, 2010).

In sum, the frequency of certain kinds of potentially damaging climate events has changed over the past half century, with additional changes in the same direction expected in the coming decade and beyond. The increasing exposure of vulnerable populations to climate and weather hazards has been the most important reason for the impacts. The exposure of people and economic assets to some climate and weather hazards (e.g., coastal storms and valley floods) is expected to continue to increase in coming decades (Intergovernmental Panel on Climate Change, 2012).

SUSCEPTIBLITY TO HARM FROM CLIMATE EVENTS

As discussed in Chapter 2, *susceptibility* refers to the likelihood of harm to a population as the result of either direct or indirect exposure to a climate event, such as a drought or hurricane. In the climate change literature susceptibility is sometimes used as a synonym for vulnerability (see Kasperson and Kasperson, 2001; Adger, 2006; Eakin and Luers, 2006; Gaillard, 2010). Adger, for example, defines vulnerability as "the state of susceptibility to harm from exposure to stresses associated with environmental and social change and from the absence of capacity to adapt" (Adger, 2006:268). We consider susceptibility as one component of vulnerability that becomes evident when or immediately after an exposed population experiences an event, and we distinguish it from actions taken following exposure to an event in order to reduce or alleviate harm, which we discuss in terms of coping, response, and recovery. We also include in our definition of susceptibility the likelihood of harm from exposure to the effects of climatic shocks and stresses that may occur in other regions, such as exposure to a spike in food or energy prices as the result of a climate-induced drought or energy supply disruption.

A large body of climate change and hazards literature explores the characteristics that influence the susceptibility of a population and its life-supporting systems to both direct and indirect harm from climatic risks and hazards (e.g., Liverman, 1990; Kasperson and Kasperson, 2001; Adger, 2006; Eakin and Luers, 2006; Leichenko and O'Brien, 2008; Adger et al., 2009b; Keskitalo, 2009; Gaillard, 2010). Factors that are widely agreed to influence susceptibility include economic, demographic, social, cultural, and environmental conditions; the form and quality of the infrastructure and the built environment; the presence of social capital; the effectiveness of institutions and governance; and the presence or a recent history of violent conflict (Intergovernmental Panel on Climate Change, 2007, 2012). These factors, briefly described below, are often correlated and interrelated. Many are also directly influenced by ongoing climatic and environmental changes (Paavola, 2008) as well as by such non-climatic processes as globalization and urbanization (O'Brien and Leichenko, 2007; Leichenko and O'Brien, 2008). Bensen and Clay (2004), for example, found that extreme storm events have long-lasting negative impacts on economic growth and development and that these effects are particularly acute in poorer regions. A study by Dell et al. (2012) found that a 1°C rise in temperature in a given year increased the probability of "irregular" leadership transitions (such as coups) in poor countries but had no effect on leadership transitions in rich countries (p. 86). Keskitalo (2009) documented the influence of the globalization of renewable resource–based industrial sectors, including forestry, fishing, and reindeer herding, on local decision making in the area of climate adaptation within Arctic communities in Finland, northern Norway, and Sweden. These examples suggest that factors influencing susceptibility are often in flux and subject to both the direct and indirect effects of other stresses (O'Brien and Leichenko, 2007).

The types of economic factors associated with increased susceptibility to harm from climate events generally include low levels of per capita income, a lack of livelihood assets and opportunities, poor functioning of local markets, and a high degree of dependency on agricultural food imports to meet basic needs (O'Brien et al., 2004; Eakin and Luers, 2006; Paavola, 2008). As discussed above, dependence on food imports can make a region susceptible to harm from systemic shocks to global food markets as the result of climatic events that affect grain-producing regions.

Key social, cultural, and demographic factors associated with increased susceptibility include low levels of education and low literacy rates within the population, high degrees of gender inequality, and large shares of elderly or dependent individuals (Intergovernmental Panel on Climate Change, 2007). High rates of population growth, particularly in urban areas, that are the result of either natural increase or immigration (see the migration discussion in Chapter 5) also increase susceptibility.

The quality of the infrastructure and the pattern and form of the built environment play a role in the susceptibility of a population to certain types of climate hazards. Within arid or semi-arid agricultural regions, the presence of irrigation and the reliability of irrigation water supply influence susceptibility to harm from drought. Within urban and coastal areas, and particularly in cities located in the developing world, poor quality and maintenance of building stock; inadequacy of water, sanitation, and energy infrastructure; and the presence of extensive areas of unplanned settlement contribute to an increased susceptibility to harm from extreme storm events and flooding (Satterthwaite et al., 2007). The condition of the housing stock in informal settlements often significantly increases the susceptibility of populations to disasters. Stories of hillsides denuded by squatter developments collapsing in the event of sudden storms or slow erosion are common. In 1975, for example, a landslide destroyed one-third of the homes in the El Agustino district of Lima, Peru; in 2000 a garbage slide in an area of Manila occupied by urban squatters killed more than 300 people and destroyed 500 homes (United Nations International Strategy for Disaster Reduction, 2009b). While a few coastal cities in high-income countries, such as London and Rotterdam, have made extensive investments in protection against coastal storm surge and sea level rise, including construction of sea walls and barriers (see London Climate Change Partnership, 2006; De Graaf and Van Der Brugge, 2010), these types of investments are currently financially infeasible for many cities in low-income countries, most of which already have significant deficits in basic water infrastructure (Parry et al., 2009).

Environmental factors affecting susceptibility to various climate-related hazards include the abundance and quality of natural assets such as forests, wetlands, and freshwater and also how well ecosystems function, which affects such things as water supply and quality, flood control, soil conservation, and biodiversity. The loss or degradation of natural assets and ecosystem services, which may occur as the result of climatic events (e.g., Carter et al., 2007), increases future susceptibility to extreme climate events of all types in different regions of the globe.

Other critical facets of susceptibility are associated with governance, institutions, and social capital. The level of public spending, the quality of the public health infrastructure, access to health care, the transparency and legitimacy of governing institutions, the presence of social networks, and the level of social cohesion all influence preparedness for extreme events and the coping, response, and recovery capacities following those events (Adger, 2003, 2006; Adger et al., 2005, 2009b; Brooks et al., 2005; O'Brien et al., 2009; Termeer et al., 2012; Wamsler and Lawson, 2012).

A final factor that influences susceptibility is the presence of conflict or political or ethnic strife. Conflict can damage the infrastructure and life-supporting systems and can undermine the capacity of institutions to prepare for and respond to climatic hazard events (Barnett, 2006; Barnett and Adger, 2007; Brklacich et al., 2010). Populations living in regions where conflict is present are highly susceptible to harm from climate risks and hazards.

Many of the above factors are generic to all types of climatic shocks and stresses, while others apply to specific types of climatic hazards, such as floods or heat waves. For example, a higher level of per capita income generally means lower susceptibility to all types of climatic stresses, while the type and quality of housing stock is more relevant to the susceptibility of coastal populations to hurricanes and other storm events. Some characteristics may increase susceptibility to specific types of climate hazards while reducing susceptibility to others. For example, within northern European and U.S. cities, brick housing stock and the lack of air conditioning have been implicated in heat wave mortality (Kovats and Hajat, 2008).

As we have already noted, many of these susceptibility factors are correlated and interrelated. A wealthier population will typically have higher levels of education, better quality building stock and infrastructure, better protection of natural assets, and more effective governing institutions, all of which reduce susceptibility. One of the main messages is the degree to which the poor suffer more and recover more slowly. As Kim (2012) recently noted:

> Globally, the poor are much more exposed to [and susceptible to the effects of] natural disasters than the non-poor, regardless of measurement methods. The poor are almost two times more exposed to natural disasters than the non-poor when measured in terms of the total number of affected people per decade. When measured by the number of disasters, the poor are 20 per cent more exposed to natural disasters than the non-poor. The time trend varies across regions, with the poor in East Asia and Pacific being most exposed to natural disasters, followed by those living in South Asia and sub-Saharan Africa. The exposure of the poor in East Asia and Pacific has started to decrease in recent years, whereas it is rising in South Asia and sub-Saharan Africa. (p. 208)

COPING, RESPONSE, AND RECOVERY

A popular but superficial image of a disaster is that a community, city, region, or even an entire nation is struck by a highly damaging event more or less evenly and that the reaction to the event is also a more or less evenly paced sequence of coping, response, and recovery. The reality is much more complex (Cannon, 1993, 1994; Wisner et al., 2004). As the United States learned with Hurricane Katrina, although the storm affected a wide swath of national territory, not everyone was affected equally, nor did everyone recover to the same level or at the same rate, for a wide variety of socioeconomic, cultural, political, and geographic reasons (Adams et al., 2006; Finch et al., 2010; Gotham and Campanella, 2011). Because susceptibilities and initial coping capacities are not evenly distributed across an affected area, loss and damage patterns are highly differentiated even within a single community, with some neighborhoods or sub-areas (or even states for that matter) devastated, but others only slightly damaged, if at all (Bankoff et al., 2004; Wisner et al., 2004; Kahn, 2005). All communities have certain stresses and problems even before a disruptive event occurs, and an event's effects will interact in various ways with those pre-existing stresses (De Sherbinin et al., 2007; Leichenko and O'Brien, 2008; Reser and Swim,

2011; Weisbecker, 2011). Box 4-1 illustrates some of these differences with the impact of cyclones in Bangladesh and Myanmar.

Highly differentiated loss and damage patterns may then be exacerbated as some parts of a community cope reasonably well with the damage and receive timely emergency assistance during response and then support for recovery, while other parts, with higher loss and damage levels and lower initial coping capacities, receive less than proportional help during

---

BOX 4-1
Cyclones in Bangladesh and Myanmar

In 1970, in what was then East Pakistan, between 300,000 and 500,000 people died from the multiple effects of Cyclone Bhola. The disaster contributed to pre-existing tensions between West Pakistan and East Pakistan and the eventual violent attempt at secession by East Pakistan, intervention by India, and the creation of the new nation of Bangladesh. With lessons learned and an innovative cyclone shelter program combined with improved public awareness, alert and warning systems, evacuation planning, hazard mitigation measures, and an understanding of local knowledge and norms, Bangladesh is now widely credited with having significantly reduced the potential for cyclone-related fatalities (Alam and Collins, 2010; Collins, 2011; Haque et al., 2012). Although no two storms are the same and all have different "signatures," making it necessary to be cautious when making comparisons, the most recent major cyclone—Sidr in 2007—killed many fewer people (only about 4,000) than Bhola, which led Haque and colleagues to conclude that there had been a 100-fold reduction compared with 1970.

In 2008 Cyclone Nargis caused the worst disaster in Myanmar's recorded history, with 130,000 dead or missing. Half of the people living in the affected areas—2.4 million out of a total of 4.7 million—were severely affected. Because the people in the region were extremely poor and had few functioning radios able to hear the only channel available, the warnings that the government issued about the impending storm were never received by most of the people there. The poor quality of the housing stock also contributed to the losses; less than 20 percent of the rural homes were able to withstand even normal monsoon rains. The storm was so strong that it overcame the local community infrastructure and social capital that had developed in large measure to compensate for the limited government presence in the region. The Myanmar government's initial rejection of numerous offers of international aid—although it eventually did allow aid into the country—almost certainly increased the loss of life. In an interview with the *New York Times* several years later, Myanmar president and former general U Thein Sein called the poor government response a "mental trigger" for moving the country from decades of military rule toward democratization (Fuller, 2012).

SOURCE: Information for these examples comes from United Nations International Strategy for Disaster Reduction (2009b) except where cited otherwise.

response and recovery. As recent works on emergency management note, transitions between disaster phases are never clearly demarcated on the ground, where it counts; response capacities vary significantly (as the U.S. federal government saw in the aftermath of Hurricane Katrina, with state capabilities in Florida versus those in Louisiana); and recovery is usually very uneven (Phillips, 2009; Coppola, 2011; Phillips et al., 2011). Developing countries are particularly prone to sharply different disaster impacts and then to sharply different coping, response, and recovery patterns, which in turn can lead to new or exacerbated social and political stresses.

In a recent article that ties susceptibilities to post-impact issues, Wamsler and Lawson (2012) note that situations are especially problematic when "poor mechanisms and structures [are] in place for response and recovery by individual residents, households and communities, or the institutions serving them" (p. 31). They emphasize location-specific vulnerabilities, which are important because disasters (except in very small nations) are almost always local or, at most, affect only a part or parts of a country directly. Thus, different effects of a disaster and then different coping, response, and recovery in affected local areas may have national effects on social and political stability, either because of the importance of the local areas or because coping, response, and recovery may reflect or influence relations among groups or regions within a country.

That the poor suffer more in disasters, both absolutely and in relative terms, has long been established (e.g., Blaikie et al., 1994; Cannon, 2000; Juneja, 2008; United Nations International Strategy for Disaster Reduction, 2009b; Kim, 2012), as has been the disproportionately severe impact of disasters on women (Agarwal, 1995; Enarson, 1998, 2012; Denton, 2002; Neumayer and Plümper, 2007; Osman-Elasha, 2009; Arora-Jonsson, 2011; Kim, 2012). Indeed, after a disaster the poor, women, and other disadvantaged groups often remain in emergency or "relief" conditions for extended periods—and sometimes permanently—while the rest of an affected community or nation moves into recovery; this confounds the simple model of disasters as following an orderly series of stages from coping to response, relief, and recovery. Thus a major lesson is that *post-impact social stresses derive not only from the total of disaster losses in a community or nation, but also—and of more concern—from how those losses are differentially experienced across groups, classes, races/ethnicities, genders, and other categories.* These stresses are exacerbated if disaster response and recovery efforts are seen as inadequate, inefficient, corrupt, or characterized by favoritism.

The steps of coping with and then responding to a disaster begin literally moments after the event starts, particularly in the most affected areas, and they are multilevel and multiactor. At the earliest stages very little coping involves the government or formal disaster response institutions.

Affected individuals, nuclear families, extended families, and neighbors in the affected communities react almost immediately, with their effectiveness largely determined by their levels of training, equipment, and social capital. Later, designated emergency response personnel arrive, with their effectiveness largely determined by their numbers, training, equipment, and logistical capabilities and support.

Initial coping is supplemented by the more formal response by various and complex combinations of emergency personnel ("first responders"); by local, subnational, and national agencies, organizations, and institutions, possibly including military organizations; and by community, religious, and other local nongovernmental organizations, and civil society in general. If disaster losses and disruptions are judged to exceed domestic coping capacity and response, the international community (bilateral, international, or national nongovernmental organizations) offers a potential additional response level. Surge capacity, the ability of assistance organizations to move needed supplies and people to the affected area in time to meet needs, is a critical factor in determining the effectiveness of the initial and early-stage response.

The coping and response levels or "waves" are not operationally distinct, because as the reaction to a hazard event deepens, the levels interact, with the optimal outcome being efficient synergy across levels and actors. The least optimal and most socially and politically damaging outcomes occur when the formal or official response conflicts with the initial coping efforts and appear to be characterized by misallocation, duplication, competitiveness, favoritism, and interagency conflict. In extreme cases, when initial coping with, and popular response to, a disaster is perceived by the leadership of the state as fundamentally threatening and is met with repression, the result can be escalating violence, as was captured classically by Cuny (1983) and recently updated and elaborated upon by Garrard-Burnett (2009) and Gawronski and Olson (2013). Thus in disaster situations, public authorities in many countries are faced with three different types of event-response tasks or challenges. First the government, broadly defined, is expected in most countries to respond using its own resources; it is also expected to cooperate with, if not support, initial popular coping efforts; and, finally, it is expected to coordinate the responses of other actors, including at times the international community. The extent of specific group and general public dissatisfaction with government disaster response or coordination is shaped both by expectations and by the perceived performance deficits in those task areas.

Given that affected and observing populations—and, in many countries, the mass media—are unusually attentive to the performance and probity of their leaders and institutions during the immediate aftermath of disasters and in the response phase, the political stakes are often quite high, including the public support or tolerance of authorities, administrations, governments, and institutions in general. In extreme cases even the legitimacy of a regime or the viability of a national society may be brought into question, a situation that can be a harbinger of possible state breakdown or failure. The processes by which social and political stresses resulting from climate events may visibly manifest in political instability or state breakdown, however, are likely to take months or even several years.

CONCLUSIONS AND RECOMMENDATIONS

The conceptual framework presented in Chapter 2 is useful for organizing available knowledge on the potential links between climate events and political and social stresses.

> **Conclusion 4.1:** *The overall risk of disruption to a society from a climate event is determined by the interplay among several factors: event severity, exposure of people or valued things, and the vulnerability of those people or things, including susceptibility to harm and the effectiveness of coping, response, and recovery. Exposure and vulnerability may pertain to the direct effects of a climate event or to effects mediated by globalized systems that support the well-being of the society.*

Because risk reflects the interactions among these factors and not only the magnitude of climate events, events of a magnitude that has not been disruptive in the past can cause major social and political disruption if exposure and susceptibility are sufficiently great and response is inadequate or is widely seen as such. The other side of this coin is that unprecedentedly large climate events do not necessarily lead to security threats if actions have been taken to reduce exposure or susceptibility or increase coping capacity and if authorities are seen to be actively responding to events.

Insights About How Climate Events Can Create Stresses

Available knowledge on the factors linking climate events to social and political stresses supports several general conclusions and points to a number of needs for further research and analysis. We note first that *each of the nonclimate factors linking climate events to social and political stresses is changing.* Many of these conditions have been changing more rapidly than climate is changing, and this situation is likely to continue for at least the next decade or two. This suggests that the net effect of climate change in the coming decades may be determined in the near term more by social, economic, and political conditions and their interactions with climate events than by climate factors alone. In particular, combinations of increased exposure and increased susceptibility are likely to have a multiplier effect on risk.

Several social, economic, and political factors that contribute to exposure and susceptibility to harm from climate events can be projected with some confidence for a decade or more at the country level or below. These socioeconomic and political factors include total population, population age distribution, level of social and economic development, land use patterns, compliance with building standards, governance capabilities, corruption levels, urbanization, certain changes in the physical infrastructure, and integration with global markets for key commodities.

Many other social, economic, and political factors that connect climate events to security threats cannot be projected with confidence at this time. However, the dynamics of climate-induced stress are well enough understood to establish several prudent expectations for anticipating climate-related security threats[1]:

1. Many societies will encounter climate-induced disruptions that stress their capacity to adapt to, cope with, or respond to the disruption.
2. The victim profile for climate events in the next decade will remain largely as it is: primarily the poor and the socially disadvantaged or marginalized. The absolute numbers of potential victims of climate events will increase with increased exposures and static or increasing susceptibility, particularly in low- to middle-income countries.
3. Harm is likely to be greatest in low- and middle-income countries characterized by high levels of corruption and weak institutions and governance, because of high susceptibilities and ineffective response.
4. Harm is likely to be greatest to populations living in countries and regions where conflict and political or ethnic strife is present or has recently been present. Such countries are more likely to have government structures that are intermediate between democratic and authoritarian and that practice inequitable allocation of public resources.
5. In some instances of ineffective response and recovery, there will be impacts on the coherence of affected states.
6. In a few instances, the consequences will be severe enough to compel international reaction.
7. If global-scale disruption does begin to occur, it is likely to appear first at especially sensitive locations and in the initial stages is likely to be interpreted as a local or regional phenomenon.

---

[1] In addition to the evidence discussed in this chapter, the research associated with the interactions of climate with some of these factors may be found Chapter 5.

8. The consequences of climate change for human societies will interact with the process of economic globalization. The known features of both processes give strong reason to expect that the conjunction of climate change and globalization will increase the risk that climate events in one location will have adverse impacts in other parts of the world.

Conclusion 4.2: *To understand how climate change may create social and political stresses with implications for U.S. national security, it is essential for the intelligence community to understand adaptation and changes in vulnerability to climate events and their consequences in places and systems of concern, including susceptibility to harm and the potential for effective coping, response, and recovery. This understanding must be integrated with an understanding of changes in the likelihoods of occurrence of climate events.*

Knowledge from several scientific fields provides useful general insights about the components of vulnerability and how they shape the effects of climate events on social and political systems. Much remains to be done, however, to advance this knowledge and to make it operational for assessing the risks of climate change to social and political systems in particular places.

A Strategy for Advancing Vulnerability Research

The intelligence and national security communities are not the only parts of the U.S. government that need to improve understanding of vulnerabilities to climate change in order to achieve national goals, and the U.S. government is not the only actor that needs improved understanding. Such improved understanding is among the objectives of the many federal scientific agencies concerned with climate change and will be valuable to the various federal, state, local, private-sector, and international organizations concerned with improving adaptation to climate change, reducing potential damage from climate events, and exploiting potential opportunities related to climate change. These shared needs for knowledge suggest that knowledge development is best pursued as a cooperative activity involving many organizations.

A recent report of the Defense Science Board (Defense Science Board, 2011) emphasized the need for federal interagency cooperation in dealing with issues of adaptation to climate change. The report notes the need for sustained attention by many federal agencies to "assisting vulnerable regions in adapting to climate change" (Defense Science Board, 2011:xiv) and calls for "a structure and process for coordination to more effectively leverage the efforts to address global problems" (pp. xiv–xv). It recommends that "the President's National Security Advisor, in conjunction with the Council on Environmental Quality, should establish an interagency working group to develop . . . a whole of government approach on regional climate change adaptation with a focus on promoting climate change resilience and maintaining regional stability" (p. vxii). The report emphasizes the need for information systems, including the translation of information into societal benefit metrics.

The analysis in this chapter clearly indicates that effective U.S. government efforts to facilitate adaptation to climate change in important regions will require knowledge about changing regional vulnerabilities as well as about climate trends. Developing fundamental knowledge about climate vulnerabilities is a major objective of the U.S. Global Change Research Program (USGCRP), which leads federal efforts to develop scientific understanding of climate change and its implications for humanity. One of the five scientific objectives in the USGCRP's strategic plan for 2012–2021 is to "[a]dvance understanding of the vulnerability and resilience of integrated human–natural systems and enhance the usability of scientific knowledge in supporting responses to global change" (U.S. Global Change Research Program, 2012:29). The intelligence community is an obvious potential beneficiary of this effort.

The USGCRP, however, faces significant challenges in advancing this research area, as noted in a recent review of the strategic plan by the National Research Council (National Research Council, 2012a). These include expanding it within a declining budget and dealing with the limited capacity of many USGCRP agencies to integrate the social sciences with climate science. Historically, the USGCRP has devoted the vast majority of its resources to understanding climate processes and only a very small portion to understanding the "human dimensions" of climate change, including vulnerability and response to disruptive climate events. This weakness of the program has been identified repeatedly in program reviews by the National Research Council (e.g., National Research Council, 1992, 1999, 2009), but the challenge remains. It might be addressed in part by improved collaboration between the USGCRP and agencies in the intelligence and national security communities that have not previously been engaged in its efforts in the domains of vulnerability and adaptation but that need the knowledge that such efforts could provide.

Conclusion 4.3: *Many of the scientific needs of the intelligence community regarding climate change adaptation and vulnerability are congruent with those of the USGCRP and various individual federal agencies. Intelligence agencies and the USGCRP can benefit by joining forces in appropriate ways to advance needed knowledge of vulnerability and* *adaptation to climate change and of the potential of climate change to create social and political stresses.*

A whole-of-government approach to understanding adaptation and vulnerability to climate change can advance the objectives of multiple agencies, avoid duplication of effort, and make better use of scarce resources.

Recommendation 4.1: *The intelligence community should participate in a whole-of-government effort to inform choices about adapting to and reducing vulnerability to climate change. It should, along with the USGCRP and other relevant science and mission agencies, develop priorities for research on climate vulnerability and adaptation and consider strategies for providing appropriate research support. The interagency effort on vulnerability and adaptation should include agencies responsible for community resilience and disaster preparedness and response domestically and internationally.*

Engagement of the security and intelligence communities could bring considerable additional resources to this effort. Establishing such an interagency process does not imply that climate change should be defined as a security issue. Rather, it indicates that security issues are among those that should be considered in developing and executing a research agenda on climate change adaptation and vulnerability.

5

Climate Events and National Security Outcomes

Traditionally, the primary security concerns of the United States and other nations have included the prevention of external assault, the prevention of insurrections and other large-scale domestic violence, and the maintenance of the political and economic stability of the state. U.S. national security concerns also extend to similar threats faced by our allies and by other states considered to be of critical importance for our national security. Other situations, such as major humanitarian crises, pandemics, or disruptive migration, which may threaten the stability of U.S. allies or other states and perhaps lead to a direct U.S. response, are also increasingly considered part of the landscape of potential security risks. All of these risks, which we refer to in the conceptual framework in Chapter 2 as security outcomes, have the potential to be affected by climate change and climate events.

Chapter 1 provided a brief review of the major scenarios linking climate change to U.S. security interests that have emerged in studies from the policy community and the intelligence and security agencies of the U.S. government. As noted there and discussed further in Chapter 2, the main focus of this report is scenarios in which climate events cause harm to systems that support human well-being by exceeding the ability of these systems to cope, respond, and recover. Depending on other factors, such harm may result in large-scale political and social outcomes that have the potential to affect U.S. national security. With the exception of events such as direct damage to military facilities caused by extreme weather, we believe the causal relationship between climate change and specific climate events and security outcomes is likely to be indirect, with complex and contingent causal pathways in between. This chapter is intended to explore the evidence that the social sciences can add to our understanding of the connections between climate events and major security-relevant outcomes. It is not possible in this study to examine all possible links, so this chapter examines a selection of some of the most commonly mentioned relationships. It begins with an examination of the connections between climate events and some of the major outcomes—such as threats to water, food, and health security; humanitarian crises; and disruptive migration—that are frequently cited in the policy literature, and it then discusses traditional security outcomes, such as political instability and interstate and internal conflict.

WATER, FOOD, AND HEALTH SECURITY

Water Security

Basics of Supply and Demand

The fundamental role that water plays in sustaining and supporting life, a healthy environment, and human well-being is drawing high-level international attention to the availability and quality of water as an essential component of development. Increasing access to safe drinking water and basic sanitation is a key component of the Millennium Development Goals (United Nations, 2012), and the period 2005–2015 was named as the United Nations International Action Decade "Water for Life."[1] More fundamentally, water is essentially irreplaceable. With other resources, such as energy and food resources, there are a number of substitutes that can be used to meet the societal needs for these resources. Currently, however, water can only be replenished at costs that are beyond the reach of many of the most water-stressed countries. Conflict over water availability or caused by issues related to delivery of water resources to meet competing needs of energy, food, and health thus have the potential to define critical climate-related conflicts and relief challenges across the globe.

Projections of future availability of freshwater suggest increasing imbalances between supply and demand. Between 1970 and the mid-1990s the amount of economically available water per person dropped by more than 35 percent (United Nations, 1997, quoted in Wolf, 2007:242), and one frequently quoted estimate (2030 Water Resources Group, 2009) projects a gap of 40 percent between global water requirements and accessible, reliable water supply by 2030. "This global figure is really the aggregation of a very large number of local gaps, some of which show an even worse situa-

---

[1] See http://www.un.org/waterforlifedecade/ (accessed July 5, 2012).

tion: one-third of the population, concentrated in developing countries, will live in basins where this deficit is larger than 50 percent" (p. 5). However, it is important to keep in mind that the models contain many uncertainties and assumptions about factors such as patterns of economic growth or the potential of technology to improve resource management, and these uncertainties and assumptions can have a significant effect on the models' results.

The agricultural sector is currently responsible for around 70 percent of freshwater consumption. Patterns of land use, population growth, and rapid urbanization, along with economic development that may require more water (for power generation or production processes, for example) and changing dietary patterns with impacts on agricultural production that also increase need for water, can be expected to have significant effects on demand, in some cases creating or exacerbating competition for supplies.[2] Contributing to the pressure is the fact that many countries depend on water sources that must be shared. As Wolf (2007) noted, "There are 263 rivers around the world that cross the boundaries of two or more nations" (p. 245). In total, these river basins account for just under half of Earth's land area, are home to 40 percent of the world's population, and make up some part of 145 countries (Wolf et al., 1999). A number of these basins—the Indus, Nile, Tigris–Euphrates, Jordan, Brahmaputra, and Amu Darya river systems, for example—are in areas of strategic importance for the United States (Office of the Director of National Intelligence, 2012). "In addition, about 2 billion people worldwide depend on groundwater, which includes approximately 300 transboundary aquifer systems" (United Nations–Water, 2008:1). Even in the absence of climate change there are multiple reasons for the intelligence community to pay attention to water issues.

---

[2] "The drivers of this resource challenge are fundamentally tied to economic growth and development. Agriculture accounts for approximately 3,100 billion m^3, or 71 percent of global water withdrawals today, and without efficiency gains will increase to 4,500 billion m^3 by 2030 (a slight decline to 65 percent of global water withdrawals). The water challenge is therefore closely tied to food provision and trade. Centers of agricultural demand, also where some of the poorest subsistence farmers live, are primarily in India (projected withdrawals of 1,195 billion m^3 in 2030), sub-Saharan Africa (820 billion m^3), and China (420 billion m^3). Industrial withdrawals account for 16 percent of today's global demand, growing to a projected 22 percent in 2030. The growth will come primarily from China (where industrial water demand in 2030 is projected at 265 billion m^3, driven mainly by power generation), which alone accounts for 40 percent of the additional industrial demand worldwide. Demand for water for domestic use will decrease as a percentage of total, from 14 percent today to 12 percent in 2030, although it will grow in specific basins, especially in emerging markets" (2030 Water Resources Group, 2009:6).

Potential Effects of Climate Change[3]

Climate change is likely to have a number of effects on water supplies, which will vary considerably across and within regions. For example, during the past several decades there have been noticeable shifts in the frequency and distribution of precipitation. Dry areas are expected to get drier and wet areas wetter. Scientists project that the subtropics, where one finds most of the world's deserts, will experience a 5 to 10 percent reduction in precipitation for each degree of global warming. Subpolar and polar regions, on the other hand, are likely to experience more precipitation, especially in the winter.

In addition, warmer temperatures mean more evaporation; warmer air can also hold more water vapor, leading to a measurable increase in the intensity of precipitation in some areas. Observations from many parts of the world indicate that a statistically significant increase in the intensity of heavy rainstorms has occurred. One of the effects of this escalation is an increased risk of flooding. And the intensity is projected to increase even in areas when overall precipitation declines.

These changes in precipitation will have a direct effect on annual *streamflow*, which is essentially equivalent to runoff, the amount of snow or rain that flows into rivers and streams. This is a key measure of the availability of freshwater. Climate models project that streamflow will decrease in many temperate river basins, especially in arid and semiarid regions. As discussed in the next section, a key question is whether the effects of climate change on water supply, combined with significant human impacts on supply and demand, could lead to tensions and conflict that become concerns for U.S. security.

Water and Conflict

Disputes over water date back millennia; the Water Conflict Chronology List, for example, begins with an account of a Sumerian legend from 3,000 BCE that resembles the Biblical story of Noah. Five hundred years later two Sumerian city-states, Lagash and Umma, provided the first written record of going to war over water; the rulers of Lagash diverted water from boundary canals to deny supplies to neighboring Umma, setting the conflict in motion.[4]

The idea that water scarcity could be a direct source of violent internal or international conflict has produced a literature on "water wars" from both academic and policy sources (see, for example, Cooley, 1984; Starr,

---

[3]The material in this section is taken from National Research Council (2012b:23–25).
[4]Water Conflict Chronology List: see http://www.worldwater.org/conflict/list/ (accessed June 20, 2012).

1991; Bulloch and Darwish, 1993; Homer-Dixon, 1994, 1996; Remans, 1995; Amery, 2002). Another literature counters that the dynamics are more complex and offer the prospect for cooperation in the management of shared resources (Elhance, 1999; Marty, 2001; Chatterji et al., 2002; Wolf et al., 2003). Concerns about water insecurity as a source of tension and conflict feature prominently in many of the government and policy community studies of the ways in which climate change could affect U.S. and international security in the coming decades (see, for example, Fingar, 2008; Defense Science Board, 2011; Office of the Director of National Intelligence, 2012). It is important to note that none of the major reports forecasts that the conflicts arising between countries over water will lead to war, although most see the potential for various forms of internal violence.

Fortunately, water is one resource for which there is a substantial research base as well as significant data sources with which to assess associations and causal linkages. One of the best known is the Water Conflict Chronology of the Pacific Institute, with data on cases from 3000 BCE to 2010.[5] Another is the International Water Events Database maintained by the Institute for Water and Watersheds at Oregon State University, which coded events from media sources between 1950 and 2008 on a 14-point scale to capture a range of conflict and cooperation behaviors.[6] Studies that examine data from both sources suggest:

- Cooperation rather than conflict is the norm with regard to water relations. In the vast majority of cases water resources are shared in a cooperative fashion and conflicts are worked out via treaties. Co-

---

[5] Current, sometimes overlapping categories of types of conflicts now include

- *Control of water resources* (state and non-state actors): where water supplies or access to water is at the root of tensions.
- *Military tool* (state actors): where water resources, or water systems themselves, are used by a nation or state as a weapon during a military action.
- *Political tool* (state and non-state actors): where water resources, or water systems themselves, are used by a nation, state, or non-state actor for a political goal.
- *Terrorism* (non-state actors): where water resources, or water systems, are either targets or tools of violence or coercion by non-state actors.
- *Military target* (state actors): where water resource systems are targets of military actions by nations or states.
- *Development disputes* (state and non-state actors): where water resources or water systems are a major source of contention and dispute in the context of economic and social development. (Water Conflict Chronology List: see http://www.worldwater.org/conflict.html (accessed June 23, 2012)

[6] The scale ranges from −7 for a formal declaration of war to +7 for a decision on unification into one nation; an international water treaty is considered a +6. For more information, see http://www.transboundarywaters.orst.edu/database/interwatereventdata.html (accessed November 15, 2012).

operative items in the Water Events Database represent two-thirds of the total over a period of more than 50 years; there are no cases in which conflicts over water lead to formal declarations of war (the most extreme form of conflict behavior on the scale) (Michel, 2009). The cases in the Water Conflict Chronology yield only one genuine interstate water war in history; in many cases water is a tool or a target rather than a cause of conflict (Wolf, 2007).

- For most of the water resources shared across national boundaries, the patterns reflect a mix of conflict and cooperation (Wolf, 2007; Zeitoun and Mirumachi, 2008).
- Most shared water resources are governed by some sort of international agreement; more than 150 international treaties to govern fresh water were put in place between 1946 and 1999 (Yoffe et al., 2003).
- Negotiations over water management and formal water agreements tend to continue even during periods of intense, sometimes violent, political conflict, including for rivers such as the Indus between India and Pakistan, the Mekong, and the Jordan between Israel and Jordan (Wolf, 2007).
- These encouraging trends aside, the cases in the Water Conflict Chronology, as well as a number of country or regional studies, show substantial conflict, some of it violent, at the national and subnational level (Postel, 1999; Wolf, 2007; National Research Council, 2012c). For example, some research suggests that as one moves from the international to the local level, the likelihood and intensity of violence increases (Giordano et al., 2002). This speaks to the importance of national political capacity, including water management systems and institutions, as well as to particular sources of local stress.

If the implications of the research on historical patterns in water resources and conflict suggest that cooperation or a mix of cooperation and conflict is the more likely outcome and that traditional interstate war is highly unlikely, how much does this tell us about the future? Should we assume that past positive trends will continue? A report from the National Research Council (National Research Council, 2012c) and four articles in the 2012 special issue of the *Journal of Peace Research* all examine water and issues of international cooperation and conflict in areas of interest to the United States. The National Research Council report examines the potential impacts of climate change on water security in the Hindu–Kush Himalayan region, which includes parts of Afghanistan, Bangladesh, Bhutan, China, India, Nepal, and Pakistan, and is the source of many of Asia's major rivers, including the Indus, Ganges, and Brahmaputra. It concludes that

Changes in the availability of water resources may play an increasing role in political tensions, especially if existing water management institutions do not evolve to take better account of the social, economic, and ecological complexities in the region. Agreements will likely reflect existing political relations more than optimal management strategies. The most dangerous situation to monitor for is a combination of state fragility (encompassing, e.g., recent violent conflict, obstacles to economic development, and weak management institutions) and high water stress. (National Research Council, 2012c, pp. 4–5)

Focusing on the inherently fragile Middle East and the Israeli–Palestinian case in particular, Feitelson et al. (2012) argue that because of increasing desalinization and water recycling efforts, climate change will have limited direct effects in that region—with the exception of Gaza, which already suffers a water deficit compounded by the ongoing Israeli–Palestinian tensions. From the authors' perspective the greater danger is if climate change generally, and water scarcity issues specifically, are taken and "used" by the contending parties to harden their negotiating positions. In an especially volatile region, that is a troubling possibility.

Bernauer and Siegfried (2012) focus on possible water conflict in the Syr Darya river basin in Central Asia, which they described as a zone that is "highly conflict-prone and [where] attempts to solve the problem have thus far failed [and where] climate change will exacerbate the problem" (p. 228). While emphasizing the possibility of increasing stress between Kyrgyzstan and Uzbekistan over runoff control and the lack of an international water management institution capable of resolving conflicts in the catchment basin, they conclude that it is more of a medium- to long-term problem. In the larger context of transboundary rivers where climate change will logically lead to "international tensions and increase the possibility of military conflict" (p. 223), Tir and Stinnett (2012) find that between 1950 and 2000 institutionalized agreements were able to offset the risks of conflict for parties to river treaties.

Taking this approach to another level and examining the recent history in 276 international river basins, De Stefano et al. (2012) are cautiously optimistic about the role of river basin organizations in "assuaging potential interstate conflict or country grievances, which may be caused by an increase in interannual water variability due to climate change" (p. 203). Looking ahead, however, they identify 14 high-risk transboundary basins with significant risks from climate change impacts. They conclude

> The picture portrayed by these data is two-fold. First, there are those well-known basins that are currently at high risk, such as the Congo/Zaire, the Niger, and Lake Chad. Secondly, there are basins with a medium present variability that are projected to experience substantial increases in vari-

ability, such as the Catatumbo basin shared by Venezuela and Colombia. Some of the BCUs [basin country units] in this latter group have very high population densities, such as the Turkish portion of the Asi/Orontes (101 people/square km), which could exacerbate the human impacts of climate change. It is interesting to note that, with two exceptions, all the basins identified as meriting further study due to present variability (eight in total) are in Africa. Conversely, by 2050, only half of the basins identified are in Africa, the rest being distributed between Latin America and Eastern Europe/Western Asia. (p. 202)

De Stefano et al. conclude that most of the current high-risk catchment basins are currently in North Africa or sub-Saharan Africa but that this will change in coming decades, with high-risk transboundary river basins developing in many other world regions.

Famine and Severe Food Insecurity

The international humanitarian community has a strict definition of "famine" along with set criteria for its declaration: At least 20 percent of households in an area face extreme food shortages with a limited ability to cope; acute malnutrition rates exceed 30 percent; and [attributable] death rates exceed two persons per day per 10,000 population. With global food production outpacing even global population growth in the past half-century, famine thus defined is no longer the specter it once was (Ó Gráda, 2009, 2011).

Occurrences of famine in the past 30 years can be attributed to access or "entitlement" issues usually associated with nondemocratic systems and price-versus-family-resource problems (Sen, 1981, 1999). Only five such events in the past three decades have been internationally declared: Ethiopia in 1984–1985, Somalia in 1991–1992, North Korea in 1996, the Gode–Somali region of Ethiopia in 2000, and Sudan in 2008. In all of these cases famine was caused either by supply-disrupting violence or interdiction and resulting isolation or else a regime's commitment to autarky (Ó Gráda, 2011).

The case of Ethiopia in 1984–1985, in which the famine had multiple causes, is particularly instructive. The sequence of events leading to the famine began with a drought, which was followed by the ruling regime's attempt to impose a socialist model of development against an ethnically and regionally based internal opposition and then the regime's diverting of international food assistance intended for that region in order to weaken the opposition. Along with the collapse of the Soviet Union, which at the time was providing the Ethiopian government with major assistance and political support, all of these factors contributed to the eventual 1991 collapse of the Mengistu regime (Keller, 1992).

Unless climate change leads to the collapse of a major global or regional environmental system and negatively affects global food stocks in a major way, in our judgment it is unlikely that famine by the above strict definition will occur in the next 10 years. There are several reasons: (1) more numerous and more consolidated democratic systems; (2) vastly improved information flows and a well-developed international monitoring and alert system (most notably the Famine Early Warning Systems Network [FEWS NET]; see Appendix E); (3) a globalized relief system with relatively fast and flexible transportation options; (4) an attentive and globalized media; and (5) a large number of proactive nongovernmental organizations with standing links to the media.

On the other hand, Ó Gráda's (2011) recent arguments on "those factors that would affect the likelihood of famine over the next decade or two" paint a much less optimistic picture, with the observation that "while democracy may prevent famine, democracy is less likely, and less likely to last . . . where famine is a risk" (p. 58). That is, special attention must be paid to countries that are either still authoritarian or where democratic systems are relatively weak and unconsolidated. Moreover, the picture—and the number of countries meriting special monitoring—becomes more complicated if we relax the strict definition of famine.

Each year the United Nations Food and Agricultural Organization places a set of countries into three at-risk tiers on the basis of their food security: those facing "exceptional shortfalls" (six in 2011), those suffering "widespread lack of access" (also six in 2011), and those facing "severe localized food insecurity" (18 in 2010). Market forces and the globalized international relief system can be expected to help these situations somewhat, but those nations in these three tiers that are also characterized by internal violence or supply interdictions warrant particular attention, especially if they are authoritarian or only weakly democratic.

Finally, because *access* to food is the crucial concept underlying entitlement and in most places *price* determines access or lack thereof, and because any real or perceived food supply problem will affect price, climate change impacts are likely to be a factor in—or blamed for—food price spikes and food security crises.

Pandemics and Health Security

An *epidemic* occurs when the number of cases of a particular disease substantially exceeds what is expected in a specified population over a given time period, and a *pandemic* is defined as an epidemic of infectious disease that has spread through human populations across large regions.

The infectious diseases with the greatest potential to cause epidemics are generally transmitted from human to human directly (e.g., influenza) or indirectly through disease vectors (e.g., yellow fever). As recent experiences with SARS and H5N1 have shown, even just the threat of a pandemic can severely disrupt business activity, trade, and travel as well as creating diplomatic challenges between countries.

Only certain diseases have the potential to cause a pandemic. The International Health Regulations, which went into force in June 2007, are intended to help the international community prevent and respond to acute public health risks that have the potential to cross borders and threaten people worldwide. The regulations provide a foundation for assessment and notification used in determining whether a public health emergency of international significance is likely. At the level of individual countries, many national governments identify diseases that must be reported by health care providers. These "notifiable diseases" are ones for which regular, frequent, and timely information regarding individual cases is considered necessary for disease prevention and control.

Any health outcome that is seasonal or sensitive to weather could be affected by climate change. This includes, but is not limited to, health outcomes associated with extreme weather and climate events, changes in air quality, infectious diseases, and malnutrition (Confalonieri et al., 2007). The health outcomes with the greatest potential to affect political stability are those in which extreme weather and climate events cause significant morbidity and mortality, leading to calls for international assistance, as well as those associated with pandemics.

The causal chain between an exposure to a pathogen and the development of a disease is complex; exposure is necessary but not sufficient to cause disease (see Figure 5-1). Infectious doses vary across diseases: Only a few pathogens are needed to cause viral diseases, while hundreds to thousands are required to cause some diarrheal diseases, such as cholera. The immune status of the exposed individual is critically important, with individuals who are undernourished, immune compromised, or suffering from other diseases generally more susceptible and more seriously affected. Furthermore, how individuals respond to an infection varies. Acute infections range from a self-limiting disease to fatal; chronic infections may leave some individuals as carriers who continue to be infectious. Other important factors influencing the burden of infectious diseases include sanitation, the quality and accessibility of public health and health care services, land use changes, population density, and travel patterns.

Emerging and re-emerging diseases are of growing concern to the public health community. These include newly recognized microbes or disease syndromes; diseases that are becoming more severe or harder to treat successfully (e.g., malaria due to drug resistance); diseases whose incidence

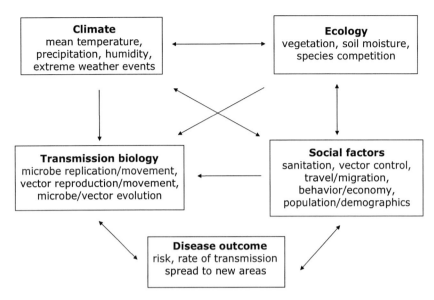

FIGURE 5-1 Causal relationships linking climate change to disease outcomes.
SOURCE: National Research Council (2001).

is increasing in areas where the pathogen is already present; and diseases expanding into areas where they were not previously present (e.g., Lyme disease in Canada). The numbers of emerging and re-emerging diseases are increasing, with 175 human infectious diseases considered to be emerging as of a decade ago (Taylor et al., 2001). The number of emerging diseases is expected to continue to increase in the future, with pathogens that infect more than one host species more likely to emerge than single-host species (Taylor et al., 2001).

Weather and climate changes have the potential of interacting with other factors to alter both the geographic range and the intensity of transmission of a number of infectious diseases, thereby creating the potential for pandemics. Evidence indicates that weather and seasonal to interannual climate variability influence the geographic distributions and seasonal variation patterns of many infectious diseases (National Research Council, 2001). Temperature, precipitation, and humidity affect the life cycles of many pathogens and vectors, thus affecting the timing and intensity of outbreaks. These variables can affect vector survival, reproduction, development, and biting rates as well as pathogen reproduction and development. Climate changes can also alter ecosystems and create other stresses in ways that influence pathogen genetics or establish new interactions between hosts

Influenza

Unlike the familiar seasonal influenza epidemics, influenza pandemics occur irregularly and spread worldwide, infecting a large proportion of the human population and causing significant morbidity and mortality. Annual influenza epidemics result from a gradual shifting of surface antigens on the influenza virus that allows it to evade host immune responses (Smith et al., 2004). Influenza pandemics can occur when a novel virus to which humans have little or no immunity, such as influenza A/H1N1, jumps to humans from avian or mammalian hosts; pigs, chickens, and ducks are the species most often implicated in this process (Centers for Disease Control and Prevention, 2009). East Asia and Southeast Asia are often the source of new influenza strains.

The best known influenza pandemic, the Spanish flu, occurred in 1918–1919 (Barry, 2005). Most of those who died were young, healthy adults between the ages of 15 and 44 (Patterson and Pyle, 1991). Because this pandemic occurred during and immediately after World War I, it is difficult to accurately estimate the total number of deaths. Mortality rates varied between and within countries and continents, with mortality in Europe estimated at approximately 1.1 percent (Ansart et al., 2009). Estimates for the total number of deaths range from more than 20 million (Barry, 2005) to 39.3 million (Patterson and Pyle, 1991) and higher.

Because influenza is strongly seasonal and the well-understood person-to-person transmission of influenza cannot explain the appearance of an epidemic, there is growing research interest in environmental triggers. In temperate regions, weather and climate variability can affect influenza incidence in two potentially complementary ways. Influenza consistently peaks during winter, when conditions for aerosol-borne transmission are favored and indoor crowding facilitates transmission (du Prel et al., 2009), and absolute humidity strongly influences the airborne survival and transmission of influenza (Shaman and Kohn, 2009). These factors provide evidence suggesting why epidemics occur during cold and dry weather. This is supported by analyses in the United States showing that the onset of wintertime influenza-related mortality is associated with anomalously low absolute humidity in the prior weeks; indeed, seasonality could be modeled with absolute humidity alone (Shaman et al., 2010). Modeling influenza-like illnesses in Europe has demonstrated a significant correlation between absolute humidity and temperature at the time of infection (van Noort et al., 2011). When a weather-dependent influenza-like illness factor was included in the model, the size of an epidemic was found to depend not only on the susceptibility of the population at the beginning of the influenza season but also on the weather conditions as the epidemic unfolded. This is consistent with modeling of the timing of pandemic influenza outbreaks, which appear to be driven by a combination of absolute humidity conditions, levels of susceptibility, and changes in population-mixing and contact rates (Shaman et al., 2011).

The El Niño–Southern Oscillation (ENSO) cycle has been implicated in influenza transmission (Viboud et al., 2004; Shaman and Lipsitch, 2012). The four most recent human influenza pandemics (1918, 1957, 1968, and 2009) were first identified in boreal spring or summer and were preceded by La Niña conditions in the equatorial Pacific (Shaman and Lipsitch, 2012). Changes in the phase of the ENSO alter the migration, stopover time, fitness, and interspecies mixing of migratory birds, which may affect their mixing with domestic animals. La Niña conditions appear to bring divergent influenza subtypes together in some parts of the world, favoring the reassortment of influenza (i.e., the mixing of the virus's genetic material into new combinations) through simultaneous multiple infections of individual hosts that can lead to the generation of novel pandemic strains.

Recent evidence indicates that dust can transport influenza viruses for long distances (Chen et al., 2010), suggesting that increased desertification from climate change and other factors could facilitate the spread of pathogenic viruses. Influenza in the tropics exhibits highly variable transmission patterns across countries, with recent studies identifying a range of environmental variables, specific to particular locations, which are associated with outbreaks (Soebiyanto et al., 2010).

Yellow Fever

Yellow fever, a viral hemorrhagic fever, has been one of the great scourges of mankind, periodically causing high mortality over at least the past 400 years (Weaver and Reisen, 2010). Approximately 200,000 cases, including 30,000 deaths, occur each year in Africa (90 percent of all cases) and in Central and South America (Barnett, 2007). Between the 18th and 20th centuries, epidemics occurred in many countries in the Americas, Africa, and Europe. Until the early 20th century many epidemics occurred in port cities of North and South America, associated with the importation of both the vector and the virus on sailing ships, where onboard transmission cycles occurred (Weaver and Reisen, 2010). In 1793 approximately 1 in 10 residents of Philadelphia died during an epidemic of yellow fever (Rogers et al., 2006). Mortality from yellow fever and malaria caused the failure of the French Panama Canal project in the 1880s and 1890s. The Yellow Fever Commission, founded as a consequence of excessive disease mortality during the Spanish–American War in 1898, concluded that the best way to control the disease was to control the mosquito that carried the virus. William Gorgas successfully eradicated yellow fever from Havana by destroying larval breeding sites. This strategy of source reduction was then used to reduce disease problems, finally permitting construction of the Panama Canal.

In contrast to the significant global campaigns to address malaria, another vector-borne disease that could be affected by climate change, a lack of continuing financial support for the vector-control strategies that successfully eliminated yellow fever from many regions led to reemergence of the disease (Gardner and Ryman, 2010). A dramatic resurgence of yellow fever has occurred since the 1980s in sub-Saharan Africa and South America, including the first epidemic in Kenya in 12 decades (Barnett, 2007). The largest outbreak of yellow fever in South America since the 1950s occurred in Peru in 1995, with cases also reported in Bolivia, Brazil, Colombia, Ecuador, and Peru. Approximately 2.5 billion people live within the current range of *Aedes aegypti*, the mosquito historically most responsible for spreading the virus, and must be considered at risk. The recent rapid spread in the United States and elsewhere of *Aedes albopictus*, a mosquito species that carries a number of arboviruses, including some that cause disease in humans, as well as dengue viruses in its native Asia, has also increased the risk of epidemics (Rogers et al., 2006).

The majority of persons infected with yellow fever virus experience no or only a mild illness (Gardner and Ryman, 2010). The incubation period in people who develop the disease is typically three to six days, with initial symptoms that include the sudden onset of fever, chills, severe headache, back pain, general body aches, nausea, and vomiting, fatigue, and weakness. Individuals who become symptomatic but recover can have weakness and fatigue for up to several months. In about 15 percent of cases, after a remission of hours to a day, a more severe form of the disease develops, characterized by high fever, jaundice, bleeding, and eventually shock and failure of multiple organs. Case-fatality rates vary widely, from around 20 percent to more than 50 percent (Barnett, 2007).

Yellow fever is now primarily a rural disease. In the neotropics, established permanent transmission cycles between nonhuman primates and canopy-dwelling mosquitoes present a constant threat of spillover to humans (Weaver and Reisen, 2010). The possible return of outbreaks of urban yellow fever is a serious potential public health risk in Africa and South America (Barrett and Higgs, 2007). Urban yellow fever results in large, explosive epidemics when travelers from rural areas introduce the virus into areas with high human population density. These outbreaks tend to spread outwards to cover a wide area, affecting up to 20 percent of the population with high case-fatality rates. No treatment outside of supportive care exists.

Mosquitoes capable of transmitting yellow fever exist in regions where the disease does not presently occur and in regions, such as Asia, where yellow fever has never occurred (Weaver and Reisen, 2010). It is not understood why the African strains of yellow fever have not invaded Asia, as have other distantly related viruses, or why urban epidemics of yellow fever disappeared from the Americas since the major campaign to eliminate *Aedes aegypti* began in the mid-20th century (Weaver and Reisen, 2010). Although *Aedes aegypti* has returned to more than its original range, yellow fever remains largely a zoonotic and rural disease in South America.

Weather and climate are major factors in determining (1) the geographic and temporal distribution of the mosquitoes that carry yellow fever, (2) the characteristics of their life cycles, (3) the dispersal patterns of the yellow fever virus, (4) their evolution, and (5) the efficiency with which they are transmitted to vertebrate hosts (Gould and Higgs, 2009). For example, Schaeffer et al. (2008) developed a tool to predict the abundance of two species of *Aedes* using climate data; water availability in breeding sites was considered the main environmental variable affecting the mosquito life cycle. The recent cases of chikungunya fever (carried by a mosquito species that is also capable of carrying yellow fever) in northern Italy support the concern that, in the presence of susceptible hosts, an environment that favors rapid mosquito replication with a suitable climate for yellow fever transmission may cause epidemic outbreaks of immense proportions.

HUMANITARIAN CRISES

Although they are terms widely used in the media and at times by relief agencies in funding appeals, there is no standard or internationally agreed upon definition of either "humanitarian crisis" or "humanitarian disaster." Both terms would seem to refer to situations in which a large number of people are found to be in immediate danger from marginal and deteriorating conditions such as those discussed earlier in this chapter (e.g., problems with food, water, or health security). The causes of situations referred to by these terms have ranged from genocide (e.g., Rwanda in 1994), war (e.g., northern Iraq in 1991), and civil conflict (e.g., Darfur after 2003) to such disasters as the Haitian earthquake of 2010 or Cyclone Nargis, which hit Myanmar in 2008. Although the terms have been stretched, if not abused, to cover many other hazards or events (e.g., the SARS epidemic of 2002–2003), the fundamental image appears to be that standard disaster-response efforts, particularly by national governments, have partially or completely failed or are about to fail and that the international community may need to be involved, often massively.

A more limited term, with its roots in the 1970s and 1980s conflicts in southern Sudan, Mozambique, and Angola is "complex humanitarian emergency" (CHE), which refers to a situation in which violence creates large population displacements either within countries (creating "internally displaced persons") or across national borders (refugees). As Natsios explained in 1995:

> [C]omplex humanitarian emergencies are defined by five common characteristics: the deterioration or complete collapse of central government authority; ethnic or religious conflict and widespread human rights abuses; episodic food insecurity, frequently deteriorating into mass starvation; macroeconomic collapse involving hyperinflation, massive unemployment and net decreases in GNP; and mass population movements of displaced people and refugees escaping conflict or searching for food. (p. 405)

Given the general global decline in violent conflict since the early to mid-1990s discussed later in the chapter, there seems to be little likelihood, especially in the next 10 years, that climate change *per se* will generate the kinds of conflicts that lead to CHEs that require massive international community responses. More likely will be "humanitarian crises" (our preferred term) where (1) large numbers of people are in immediate peril from either a climate-related, relatively fast-onset event (e.g., a heat wave, flood, cyclone, or pandemic) or a relatively slow-onset event (e.g., a drought, crop failure, food shortage, or forced migration); (2) local coping and national responses are seen as inadequate to address these needs, thus requiring major assistance from the international community and, in particular, the U.S. government; and (3) the dynamics of the longer-term recovery may lead to security conditions of major concern to the United States.

DISRUPTIVE MIGRATION

Climate change–induced migration is often implicated in scenarios in which climate change leads to violence (Reuveny, 2007). This section considers the current state of knowledge on the connections between climate change, migration, and national security. In defining migration, a distinction is typically made between internal migration, which entails population movement within a country, and international migration, where population movement extends across international borders. It is also important to keep in mind other features of population movement, such as whether it is temporary or permanent and whether it is voluntary or forced. Even within these different categories, migration can take a variety of forms, including: temporary or permanent displacement of a population following some type of climate event or other disruptive event, such as a tsunami or nuclear accident; forced or voluntary migration out of an area of political or military conflict; temporary or permanent relocation of a population from an area threatened by flooding or inundation; and temporary or permanent movement from one region or country to another for economic opportunity. Climate change may be directly or indirectly associated with many of these different forms of migration.

It important to recognize that most types of migration are not regarded as a direct security threat and that the decision to migrate is in fact often an appropriate and effective response under a wide variety of circumstances. Given the emphasis in this report on climate change and U.S. national security, we are particularly interested in a specific type of migration, which we term "disruptive migration." Disruptive migration, which may be internal or international, generally involves large-scale movements of populations that are socially, economically, or politically disruptive, either in the area of origin, the area of destination, or in sensitive border regions that may be affected by population movements.

Some writers on the connections between migration and environmental change have characterized climate change as a major driver of migration (e.g., Warner et al., 2010). Yet the empirical evidence base on this issue is extremely limited (Black et al., 2011b; Lilleør and Van den Broeck, 2011). In a recent comprehensive review of estimates and predictions of human displacement as the result of environmental change, Gemenne (2011) finds a notable lack of consensus on either the number of individuals displaced by environmental change or a methodology to derive such a number. Some key reasons for this lack include disagreement about the definition of an "environmental migrant" or "environmental refugee," limited data on the magnitude of internal displacements, and failure in many studies to distinguish between ongoing demographic shifts and population changes that are the result of migration (Biermann and Boas, 2010; Gemenne, 2011). Gemenne (2011) further notes that most studies of this type emphasize predictions about potential future displacement or population movement based on estimates of populations that will be exposed to environmental stress such as storm surges and sea level rise (e.g., Biermann and Boas, 2010) rather than on documentation of the actual numbers of people displaced by past or ongoing environmental change.

Despite limited evidence on this issue, most studies recognize that the linkages between migration and environment are multifaceted and complex (Perch-Nielson et al., 2008; Warner, 2010; Afifi, 2011; Black et al., 2011b; Hugo, 2011; Renaud et al., 2011; Seto, 2011). In examining recent evidence on the connections between environmental change and migration, Black et al. (2011b) drew upon the broader literature on the drivers of migration and concluded that changes in environment are among a "family" of migration drivers (see Figure 5-2). These drivers, which include economic, political, social, and demographic factors, generally act in combination, so that environmental drivers cannot be entirely separated from other types of drivers (Black et al., 2011a). Seto (2011) comes to a similar conclusion

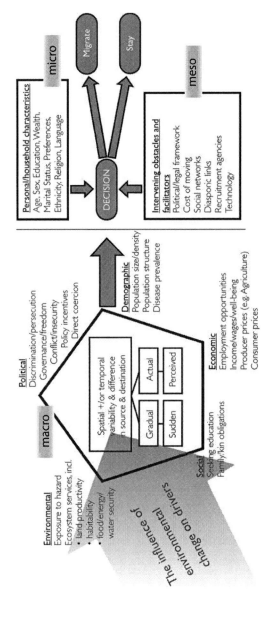

FIGURE 5-2 Environmental drivers interact both directly and indirectly with other drivers of migration.
SOURCE: Black et al. (2011b). Copyright 2011. Reprinted with permission from Elsevier.

in a meta-analysis of recent scientific literature on the drivers of migration into 11 coastal megacities located in delta regions in Asia and Africa. The study finds that migration to these cities, which include Bangkok, Dhaka, Guangzhou, and Karachi, has primarily resulted from a combination of economic, demographic, political, and social factors rather than environmental conditions. Yet the study also notes the difficulty of separating out these other factors from environmental conditions such as land degradation, land scarcity, or extreme events (Seto, 2011).

In exploring the environmental drivers of migration in more detail, Black et al. (2011b) suggest that environmental change may directly contribute to migration through mechanisms that contribute to changes in the reliability or availability of ecosystem services such as productivity of land; food, energy, and water security; and exposure to hazards. Environmental change may indirectly contribute to migration through its effects on economic drivers, such as livelihood opportunities, or on political drivers, such as conflicts associated with availability and access to resources. Connections between environmental change and migration can be seen in the case of Niger, where Afifi (2011) found that environmental degradation, including the interrelated problems of drought, deforestation, and soil degradation, led to reduced incomes among farmers, herders, and fisherman, thereby contributing to economic migration. Afifi's study (2011), which was based on fieldwork conducted during 2008 in two regions of the country (Niamey and Tillabéri), found that migrants were typically young men who left their families behind in search of work either in another region or in another country. Most of the approximately 60 migrants interviewed in the study identified economic factors such as poverty and unemployment as key reasons for migrating, but almost all (90 percent of those interviewed) indicated that environmental problems also played a role in their decision to migrate (Afifi, 2011).

Recent literature further emphasizes that individual or household migration decisions are generally dependent not only upon social, economic, political, and demographic drivers, but also on the particular characteristics of individuals and households, including level of wealth, education, worldviews, ethnicity, and so forth. Individual migration decisions are also influenced by what Black et al. (2011b) term "intervening obstacles or facilitators," such as an individual's social network, legal and political mechanisms, the presence of recruitment agencies, and so forth (see Figure 5-2). As such, the same set of "structural" drivers may contribute to different outcomes for different households depending both on a household's characteristics and on intervening barriers and facilitators to migration.

While caution is needed when generalizing from past findings to a future climate where extreme climate events will be more likely and more severe, we can postulate that disruptive migration may potentially result from direct environmental stresses associated with climate change as well as from the indirect effects of climate change–induced environmental change on economic and political conditions. Climate change may constitute a direct environmental driver of either temporary or permanent migration via its effects on the availability of ecosystem services including, for example, the supply of freshwater, which may change under altered rainfall regimes; coastal flood protection, which may be lost as the result of sea level rise; and changes in the productivity of agricultural lands as a result of changes in temperature and precipitation regimes (Black et al., 2011b). Climate change may also affect the likelihood of droughts, coastal storms, and other types of hazardous climate events, which may temporarily or permanently displace susceptible households. Climate change may indirectly contribute to migration, whether temporary or permanent, via effects on economic, political, and social drivers. For example, climate change may influence agricultural and natural resource–related livelihood opportunities in a particular region, or it may contribute to political conflicts within a region over water or other resources. In all of these cases climatic shocks and stresses interact in complex ways with the other known drivers of migration so that the effects of climate events are not monotonic (i.e., more intense climate events do not necessarily lead to more migration) and they also depend on other causal variables (Warner, 2010; Black et al., 2011b).

When considering the migration decisions of individuals and households, it is important to keep in mind that climate changes may lead to alterations in a household's characteristics, which, in turn, can influence the actions of individuals. For example, climate change may create conditions that result in a long-term decline in a household's wealth and assets, which in turn can both increase the motivation to migrate and decrease the ability of members of the household to migrate in search of new economic opportunities. Policies intended to facilitate adaptation to climate change may provide incentives for a household either to relocate or to remain in a region. Another critical factor for migration decisions is the effectiveness of responses to climate events. As noted in the discussion of humanitarian crises, effective response is a key determinant of whether an extreme climate event becomes a humanitarian crisis. Extreme events have often resulted in temporary, internal population displacement but have rarely led to permanent migration (Perch-Nielsen et al., 2008; Lilleør and Van den Broeck, 2011). Effective immediate responses to extreme events and effective recovery efforts might therefore be expected to reduce population displacement or to shorten the duration of displacement.

Several types of security threats can be identified as possible results of climate change–related migration. Broad migration trends indicate that over the past 30 years large cities have been growing faster in low-elevation coastal zones than elsewhere and that these trends are likely to continue

(Seto et al., 2011). New migrants to these cities tend to be highly exposed to climate stresses, including storm surge and sea level rise, because they often tend to move to locations, such as floodplains and hill slopes, that are highly exposed to environmental hazards. They also tend to be more susceptible to being harmed because, within receiving cities, migrants are typically poorer than resident populations and in many cases have little knowledge of local environmental risks in a new city and may therefore be unaware of the risks of moving to areas that lie within flood plains or are otherwise hazard-prone. These interrelated trends suggest that climate change–related threats to human security may be just as prominent in areas of migration destination, particularly urban ones that receive large numbers of immigrants, as in areas of emigration. Migrants into new areas may also place strains on governmental or other resources and may potentially contribute to new types of conflicts, particularly within receiving areas that are already under social stress (Reuveny, 2007). Large flows of immigrants can also create or contribute to security threats in sensitive border or transition regions.

Although much of the discussion on climate change and migration emphasizes displacement or unplanned migration, it is also important to recognize that migration is an adaptation strategy to environmental change that may be necessary and even beneficial under some circumstances (McLeman and Smit, 2006; Tacoli, 2009; Black et al., 2011c; Foresight, 2011). In some cases migration may reduce security threats by taking pressure off local resources or by bringing new revenue into a local area as the result of international remittances. Paradoxically, climate change may also undermine options for adaptation via its effects on the assets and other characteristics of households. Migration typically requires a significant outlay of financial resources, yet actions needed to cope with environmental changes (e.g., selling land or livestock) can reduce a household's assets to the point that family members who could adapt by migrating may not have the resources to do so. Those households or individuals who cannot migrate out of a region that is undergoing environmental change are among the most vulnerable (Black et al., 2011c). Regions with large concentrations of "trapped" populations that are unable to migrate may pose a new type of human security threat. When an extreme climate event occurs, these "poorest of the poor" may end up trapped in environmentally degraded areas, creating situations that are ripe for humanitarian catastrophe (Foresight, 2011).

SEVERE POLITICAL INSTABILITY AND STATE FAILURE

Extreme political instability, particularly when it substantially weakens or causes the overthrow or collapse of a strategically important regime or when it results in the onset of civil war, may have significant security

> **BOX 5-1**
> **The Political Instability Task Force**
>
> The Political Instability Task Force (PITF) is an ongoing and unclassified research program funded by the Central Intelligence Agency that began work in 1994 as the Task Force on State Failure, a panel of academic scholars and methodologists. Its original task was to assess and explain the vulnerability of states around the world to political instability and state failure, focusing on events like the collapse of state authority in Somalia and the former Zaire and other onsets of disruptive regime change, civil war, genocide and mass killing, and onsets and terminations of democratic government. The task force uses open-source data and research to develop statistical models that can accurately assess countries' prospects for major political change and can identify key risk factors of interest to U.S. policy makers.
>
> SOURCE: Personal communication with Lawrence Woocher, PITF research director.

consequences for the United States. At least since the fall of the Shah of Iran in 1979, the U.S. government, including the intelligence community, has invested heavily in research to provide the basis for understanding and predicting the sources and onset of political instability in countries and regions of concern. Following the collapse of Somalia and the war in the Balkans, another related literature on "state failure" developed, again with substantial support from the U.S. government, including the intelligence community. In addition, other governments supported comparable research, and there was active research in the academic and think tank community as well.[7] Two reviews of the literature on state failure (Bates, 2008; Marten, 2010) make clear the close relationships with efforts to understand the origins of extreme political instability, in particular, armed internal war.

One example of these connections, described in more detail in Box 5-1, is the work of the Political Instability Task Force (PITF), a research effort funded by the Central Intelligence Agency that began in the mid-1990s as the Task Force on State Failure. Over time and with continuing adjustments, the PITF has developed a model that is able to correctly identify the onset of "adverse regime change" or "ethnic or revolutionary war" two years in advance in 80 percent of the recorded instances. The methodology and results for a set of cases between 1955 and 2003 are described in the most recent published account of the PITF's research available (Goldstone

---

[7]Marten (2010) includes discussions of a number of the other major research projects on state failure.

et al., 2010), although the committee also received a briefing on its more recent work (Goldstone, 2012).

As discussed further in Chapter 6, the PITF has examined hundreds of potential explanatory variables, finally concluding that the countries most susceptible to internal violence have been partial democracies pursuing policies that favor one segment of their population over others. The other two variables in the researchers' statistical model, in addition to their characterization of political institutions and allocational policies, are infant mortality and the incidence of conflict in bordering states. The results, published in 2010, include some climate-related variables (e.g., the impact of drought) in the list of those that did not prove to be statistically significant.

Extending that assessment, Hewitt et al. (2012) developed the Peace and Conflict Instability Ledger, which assigns risk factors to all countries in the world (again without consideration of climate variables). This effort finds regional concentrations of risk in South Asia and in sub-Saharan Africa, suggesting that unusual climate events in those regions are of particular concern in terms of exacerbating the potential for political instability. As discussed further below, the Indus River valley is an area of particular concern because of a combination of political and environmental factors, including unusually severe drought and flooding, and a political process that has favored using available water for irrigation rather than for power generation but that has not allocated water equitably between the favored Punjab and the arid Sind regions. On the Peace and Conflict Instability Ledger, Pakistan has the seventh-highest risk factor in the world, with neighboring Afghanistan having the highest. In addition, while water-related issues in the Middle East, particularly among Israel, Syria, and Turkey, have been managed relatively well to date, the underlying tensions in the region could begin to affect that cooperation.

The recent literature exploring potential links between climate events and violent conflict is discussed in the next section. As described above, the literature on other forms of extreme political instability has generally not explored potential climate–security connections. As the Bates (2008) and Marten (2010) reviews make clear, most of the efforts to understand the origins of state failure focus primarily on economic factors, various forms of ethnic divisions, and the state of democratization in a particular country. Within and across each of these major categories, there is substantial disagreement on causal pathways and on which factors are likely to exert the most influence on the survival of a regime.

Bates (2008) also echoes suggestions from other literatures on the need for more data from below the national level to gain better insights into the dynamics of state failure:

> Although recent research has tended to emphasize the political wellsprings of state failure, future research needs to employ new kinds of data. In addition to incorporating information concerning deeper political forces, it needs to make systematic use of subnational data. The origins of political disorder lie in conflicts whose own origins are, to a great degree, internal to the nation-state: regional inequality, conflicting partisan preferences, religious differences, and so on. Aggregate, national-level data offer the wrong optic by which to view within-country conflict. (p. 10)

His review cites several examples of what he considers an encouraging trend of introducing this type of data in an increasing number of studies.

One literature that does provide a more detailed exploration of potential climate–security links is the literature on the potential political impacts of disasters. Its findings generally support the conclusion that climate events that trigger disasters of various types are associated with political instability, although not in a straightforward way. The relationships, including causes and effects, are highly complex and contingent. The overall analytic challenge was well captured in a recent review of detailed analyses of several major disasters of the past, including some that led to state failures (Butzer, 2012). The review found that in many, but not all, instances, states survived the calamities, and it cautioned against drawing too straight a line between disasters and state failures, noting that state breakdowns differ because of the "great tapestry of variables" involved.

Studies of the political consequences of natural disasters in the modern era provide another source of useful insight. A series of case studies of different types of disasters by Olson and various collaborators (Drury and Olson, 1998; Gawronski and Olson, 2000, 2013; Olson, 2000; Olson and Gawronski, 2003, 2010; Poggione et al., 2012) indicates that while disasters often become quite "political," disasters that result in major violence, falls of government, regime changes, and even state breakdowns are relatively rare in comparison with the total numbers of annual disasters. Indeed, the researchers' line of argument suggests that a certain amount of political unrest and even violence should be expected in post-impact disaster situations, particularly when the response appears inadequate. Extrapolating from their work, it would seem that large-scale violence, regime change, or state breakdown requires a particular combination of factors that often take months to several years to manifest visibly: (1) incumbent political authorities with little public support at the time of the disaster; (2) a disaster response that is perceived to be under-resourced and poorly managed, especially if it is seen as characterized by favoritism, corruption, and lack of compassion; (3) a regime lacking broad, value-based "diffuse" legitimacy and dependent upon "specific" legitimacy (material rewards to key groups); and (4) well-organized pre-existing opposition groups within the system (e.g., political parties) or outside it (insurgent, separatist, or revolutionary movements) that are capable of leading, organizing, and engaging in or increasing already existing anti-government or anti-regime violence.

The scenarios in which climate events are most likely to lead to risks to U.S. national security are in countries of security concern that have a significant likelihood of exposure to particular climate events combined with susceptible populations and life-supporting systems, weak response capacity, and underlying sources of potential political instability. Pakistan offers a case that illustrates these points particularly well, as described below. Another potential case, Egypt, is presented in Box 5-2.

BOX 5-2
Egypt

Of the many places in the world where climate dynamics might induce globally consequential disruption within a decade, Egypt is a principal possibility. Egypt's population of some 80 million people consumes 18 million tons of wheat annually as a dietary staple, half of which is imported, with virtually all the rest dependent on water from the Nile River. The Nile flows through Sudan and Ethiopia before entering Egypt and accumulates nearly all of its volume upstream. The production of wheat and other food crops supported by the river is being burdened by population increases in all three countries. The countries' current combined total of 208 million people is projected to reach 272 million by 2025, presumably generating an increase in agricultural production demand on the order of 30 percent or more within the watershed. In addition South Korea and Saudi Arabia have purchased large tracts of land in the watershed to assure imports for their own populations, and that will also add to the demand for water (Brown, 2011).

At the moment there is no broadly agreed projection of water flow in the Nile over the next decade and hence no widely accepted basis for estimating the risk of climate-induced disruption. There are historical reasons for acknowledging the possibility, however. In particular, between 1961 and 1964 there was a sharp increase in annual rainfall over Lake Victoria, adding a substantial amount of water to the White Nile branch of the watershed. Annual rainfall over the lake has receded in the intervening years, but it has not yet returned to the levels that generally prevailed from the time that annual records were initiated in 1869 up until 1961 (Sutcliffe, 2009). If that were to occur on a sustained basis, the three countries primarily affected would encounter very serious water management problems with no agreement in place to organize their respective responses. Their populations would encounter a threat to their food supply that no allocational arrangement would likely remove. In the past the existence of the Aswan Dam has buffered Egypt against fluctuations in river flow, but the continued increases in water demand, plus the potential loss of flow noted in the second paragraph, increase the potential that such buffering will be insufficient in the future.

Pakistan as an Example

Pakistan presents a clear example of a country where social dynamics and susceptibility to harm from climate events combine to create a potentially unstable situation. Pakistan's economy depends heavily on water from the Indus River, and competition for this water is increasing. Therefore, Pakistan's political and economic systems may be vulnerable to hydrological changes in the Indus system such as have been observed recently and which may be affected by climate change and variability at a subcontinental scale.

Agriculture is a central component of the Pakistani economy. The sector accounts for 21 percent of annual gross domestic product (the second-largest fraction by sector) and is by far the largest source of employment, employing 45 percent of Pakistani workers (Government of Pakistan, 2012). Furthermore, these percentages do not capture the dependence of other sectors on agriculture. Much of the agricultural production feeds domestic industry, particularly the cotton grown for the country's large textile industry. Textiles and clothing make up a very large portion of Pakistan's exports—approximately 50 percent in recent years—thus representing the country's most important source of foreign currency (Government of Pakistan, 2012).

Given the low levels of rainfall in the agricultural areas of the country, Pakistan's agricultural sector relies heavily on irrigation. The ratio of area of irrigated to rain-fed agricultural land is 4-to-1, the highest ratio worldwide (Nizamani et al., 1998). Water for irrigation is drawn primarily from three storage reservoirs on the Indus, making this crucial economic sector highly dependent on adequate flows in the Indus system.

Further stressing the Pakistani water system, demands for water for agricultural, domestic, and industrial uses are increasing. Agricultural production is intensifying, shifting from subsistence crops to commodity crops (mostly cotton, sugarcane, and rice) that produce more output but require more water; manufacturing activity is increasing as a share of the economy; and population growth, especially in urban areas, is requiring more withdrawals of Indus water for domestic consumption (World Bank, 2005). Additionally, hydroelectric power provides 37 percent of Pakistan's electricity (Government of Pakistan, 2012), mostly from the reservoirs also used for irrigation-water storage, creating competition for water resources between agriculture and energy, at least at some times of the year.

Competition for water between the agricultural and power sectors is already intense and is likely to increase. Policy decisions by the government, dominated by influential landowners of Punjab and Sind, have allocated water to irrigation purposes instead of to power generation (Ghumman, 2012a), triggering protests from industrialists and the general urban population. Protests over power outages, although not new in Pakistan, have led to increasing civil unrest over the past five years (News International, 2012). With the onset of a sweltering summer, power shortfall hit a record high of 8,000 megawatts in 2012, or nearly 45 percent of national demand (Ghumman, 2012b), leading to 18 to 20 hours per day of power outages and stoking riots and mass-scale protests (Ghumman, 2011a). Reports from the ground recorded violent protests throughout the country. In a recent episode of escalating violence, rioters burned trains, damaged banks and gas stations, looted shops, blocked roads, and, in some instances, targeted homes of members of the National Assembly and provincial assemblies (*Express Tribune*, 2012). According to a senior local police officer in the largest city, Karachi, on average there were at least six protests against power outages in the city per day in 2011 (Dawn, 2011). Competition between water uses is likely to increase if government plans are implemented to increase hydroelectric capacity as a cheaper alternative to imported fossil fuels (Pakistan Water and Power Development Authority, 2011).

As a result of these demographic and economic changes, an already tight water supply is becoming increasingly stressed, to the point that Pakistan has been described as "one of the most water-stressed countries in the world" (World Bank, 2005:viii). In quantitative terms, in 1951 some 6,880 cubic yards of surface water were available per person in Pakistan. By 2010 that had dropped to 1,358 cubic yards per person, and it is projected to decrease to 1,046 cubic yards by 2025 (Pakistan Water and Power Development Authority, 2011).

Recent hydrological events in the Indus system suggest the kinds of stresses climate events can put on this society. In 2010 a shift in the distribution of monsoonal rainfall led Pakistan to experience massive flooding, inundating one-fifth of the land area of the country, in one of the worst natural disasters the country has faced (National Research Council, 2012c). In 2011 the Indus River System Authority (IRSA), Pakistan's institution responsible for monitoring the flow of Indus basin rivers and apportioning its waters among the provinces, reported indications of drought and water shortage at storage reservoirs on the Indus, according to press reports (Dawn, 2011). Although individual extreme events cannot be attributed with confidence to climate change, increased frequency and intensity of drought and flooding are consistent with climate change projections.[8]

Beyond the short-term events, there is some evidence that the mass

---

[8] The Intergovernmental Panel on Climate Change fourth assessment report (Intergovernmental Panel on Climate Change, 2007) projects lower annual average precipitation in the subtropics but also a greater incidence of high precipitation events, so that a greater concentration of total precipitation is expected to occur in extreme events and an even greater increase in dry periods is expected than implied by the drop in average annual precipitation.

balance of the Karakoram glaciers in the headwaters of the Indus system—the source for the great majority of the river's water (Archer and Fowler, 2004)—has been changing in ways that may reduce river flows. Glacial and snow melt are more important to water supplies in Pakistan than they are to countries farther east in the Himalayan region, where monsoons provide a much larger share of river flows (Bolch et al., 2012). Precipitation levels in winter, when most glacial accumulation occurs in the Karakoram area, have recently increased (Archer and Fowler, 2004; Bolch et al., 2012). Moreover, in contrast to most other regions in the world, mean air temperatures in the Karakoram have decreased in summer, when most of the loss of glacial mass occurs (Bolch et al., 2012). Hewitt (2005) found that Karakoram glaciers began increasing in area in the 1990s after several decades of observed retreat, but he could not reliably assess the mass balance (as opposed to areal extent) of the glaciers with the available data. It is clear from the data that the Karakoram glaciers are behaving quite differently than the rapidly retreating glaciers of the eastern Himalaya. The Karakoram glaciers show stable or increasing areas and possibly mass (National Research Council, 2012c).

Definitive statements on changes in the glacial mass and in the river flows in the Indus basin would require data from land-based observations of the glaciers, which do not exist for lack of observing stations, and hydrological records from the Indus, which are classified by the Pakistani government. However, the available observations do fit a coherent story of changing glacial mass, which Hewitt (2005) attributes to climatic changes. Because streamflow in the Indus system is so dependent on glacial and snow melt (Archer and Fowler, 2004), and because the Pakistani economic and social systems are so dependent upon the Indus, changes in the mass balance of Karakoram glaciers or hydrology of the Indus are of extreme importance to Pakistani society.

Despite the threats that climate change may pose to the Pakistani society through effects on water supply, there is no public mechanism in place to record the fluctuations either on the upstream or the downstream parts of the Indus basin and no solid basis for making projections. The only data available to gauge the fluctuations in river flows are maintained by IRSA and are classified for political reasons. Press accounts and some technical reports indicate the country's increasingly stressed water resources (e.g., Pakistan Institute of Legislative Development and Transparency, 2011) and its leaders' difficulties dealing with the problem. For example, Pakistan's hydropower infrastructure has not been able to meet the ever-growing power demands, and the government has been unable to set a target date to solve the problem (Saifuddin, 2012).

The Pakistan case illustrates well how a highly stressed environmental system on which a tense society depends can be a source of political instability and how that source can intensify when climate events put increased stress on the system. It also illustrates the value to security analysis of monitoring the many social, economic, environmental, and political elements of such as system. We return to the monitoring issues in Chapter 6.

INTERSTATE AND INTRASTATE CONFLICT AND VIOLENCE

Patterns of Violent Conflict

As background for the discussion of research about climate–conflict connections, it is useful to note several general trends in global patterns of internal and interstate conflict since the end of World War II. Traditionally researchers have used the threshold of 1,000 battle-related deaths in a year when defining a "war." There are several large databases that track the incidence of conflicts, including different types of wars and armed conflicts around the world. In addition, there are projects to track other forms of political violence (e.g., armed attacks and political murders) or political conflict that may fall short of violence (e.g., riots).[9] And scholars frequently develop their own datasets to explore specialized topics, such as detailed explorations of patterns of conflict and political violence in a particular region. As discussed further in Appendix E, government agencies, private foundations, and others are making significant investments in efforts to take advantage of new technologies and enhanced computing power in order to gather more refined data about political and social violence of all types.

Much of the research on climate–security connections relies on the Uppsala Conflict Data Program (UCDP) at Uppsala University in Sweden, which has been collecting data from public sources since the 1970s.[10] UCDP provides data on "armed conflicts," in which at least one party is the government of a state and for which battle-related deaths range from 25 to 1,000 in a year, and "wars," in which battle-related deaths surpass 1,000 a year. The dataset goes back to 1946 and at the time of this report extends through 2011. UCDP also maintains a number of other specialized databases on such things as intrastate conflict without the involvement of a government (data

---

[9]A listing of the major datasets on international conflict and cooperation, maintained as an online appendix to the ISA Compendium Project of the International Studies Association, may be found at http://www.paulhensel.org/compendium.html.

[10]Information about UCDP, including descriptions of its databases, definitions of its variables, and a variety of graphics, is available at http://www.pcr.uu.se/research/UCDP/.

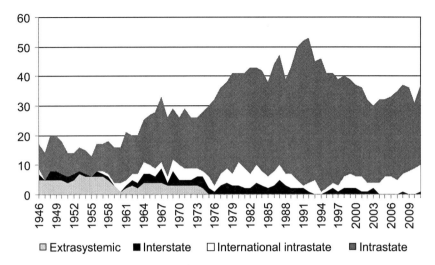

□ Extrasystemic ■ Interstate □ International intrastate ■ Intrastate

FIGURE 5-3 Trends in armed conflicts, 1946–2011.
SOURCE: Themnér and Wallensteen (2012). Copyright © 2012 by Sage Publications. Reprinted by permission of Sage. Graphic also available from the Uppsala Conflict Data Program (UCDP) website, see http://pcr.uu.se/digitalAssets/122/122554_conflict_type_2011jpg.jpg (accessed December 26, 2012).

available from 1989 to 2011). As shown in Figure 5-3,[11] the UCDP data reflect two broad trends that also appear in the other major databases:

- The decades since World War II have experienced a relatively small number of interstate armed conflicts, both in absolute numbers and, especially, relative to intrastate conflict. This is particularly noteworthy because, with the end of colonial rule and then the dissolution of the Soviet Union and other states following the end of the Cold War, the number of nation-states has grown substantially. For

---

[11] The four types of conflicts shown in the figure are defined as:

- "Interstate conflict," a conflict between two or more governments;
- "Extra-state conflicts," the term used primarily for colonial conflicts, the last of which ended in 1974;
- "Intra-state conflicts," conflicts between a government and a nongovernment party, with no interference from other governments; and
- "Internationalized intra-state conflicts," conflicts in which the government side, the opposing side, or both sides, receive troop support from other governments that actively participate in the conflict. See Uppsala Conflict Data Program, http://www.pcr.uu.se/research/ucdp/definitions/ (accessed June 2, 2012).

example, the United Nations has grown from 51 founding members in 1945 to today's 193 member states.
- After peaking in the period immediately following the end of the Cold War, the number of intrastate armed conflicts generally declined through 2010.

In 2011 the number of active conflicts increased to 37, from 31 in 2010, the largest increase between any two years since 1990 (Themnér and Wallensteen, 2012), although the total was still well below the peak of 53 active conflicts in the early post-Cold War years. Six of the conflicts were categorized as wars; the conflict between Cambodia and Thailand, although considered minor, represented a new interstate conflict. The growth "was primarily driven by an increase in conflicts on the African continent, and is only in part due to events tied to the Arab Spring, which mostly led to other forms of violence than conventional armed conflict" (Themnér and Wallensteen, 2012:1).

One implication that the small number of interstate conflicts over the past 60 years has for research on climate–security connections is that it is difficult to use the limited recent experience to forecast potential trends. Other issues related to interstate wars and crises are discussed below.

Interstate Wars and Crises

The limitations imposed on forecasting by the relatively small number of interstate wars in recent decades are compounded by the continuing changes in the fundamental characteristics of the international system since the end of the Cold War. These circumstances make it extremely difficult to test competing hypotheses about risk factors for interstate conflict that would be relevant to current circumstances. In addition to these difficulties there is a lack of consensus among scholars about the causes of such wars (Levy and Thompson, 2010) and about how they compare with the sources of internal conflict. These are problems that affect any effort to understand the risks of a return to more frequent interstate conflict.

On the specific question of climate change as a source of interstate conflict, there is very little systematic evidence, and the causal mechanisms are not well understood. One study (Tol and Wagner, 2010) did find that, for the period 1000 to 1900, low temperatures in Europe coincided with an elevated risk of interstate war. Beyond that finding, prolonged severe drought is the most common type of correlation studied, but this may be in part because it is the easiest anomaly to detect in the ancient record, and it may not be representative of the consequences of climate change in general. As discussed earlier in the chapter, the one area in which there is clear historical evidence of an effect—disputes over shared water resources—the evidence indicates that the outcome is more likely to be cooperation than war (Wolf, 2007). As we already noted with regard to conflicts over water, the literature from the policy community and government agencies discussed in Chapter 1 generally does not foresee that the effects of climate change will lead to war in the traditional sense of violent interstate conflict. Other forms of conflict—including potential violence within states—are considered far more likely.

During the Cold War the possibility of a regional political crisis spiraling out of control and leading to direct U.S.–Soviet confrontation led to substantial research on the dynamics of interstate crises (George, 1991; McCalla, 1992). Much of the recent research, however, has focused on the risks of crises that lead to the start or recurrence of internal conflicts.[12] There has also been almost no effort to explore empirically whether climate factors might lead to or exacerbate tensions between states to a point short of outright war. Again, disputes over water resources are the one exception. A better understanding of how climate change or events might affect crisis dynamics, particularly in regions where there are other reasons to be concerned about the risks of interstate conflict, could contribute to an understanding of the potential for violence.

Internal Conflict

This section reviews the new and rapidly growing academic literature that explores links between climate stress and internal armed conflict. In the 1990s the accumulating evidence and emerging scientific consensus that the planet was in the early stages of a fundamental and profound climate change was accompanied by suggestions in the peer-reviewed academic literature that this change could lead to increasing levels of violence. The link between climate stress and the outbreak of internal war has been the subject of a dramatic increase in original empirical research about climate–security connections (see Figure 5-4).

The core thesis for those arguing for a link between climate and violent conflict is that climate change–induced health problems and resource scarcity (in particular, the availabilities of water, food, and energy) will lead to interstate violence and intrastate unrest, instability, and armed conflict in the most directly affected nations or regions. Homer-Dixon (1991, 1994, 1999, 2007) and Swart (1996) were among the earlier articulators of this concern in the peer-reviewed literature, followed later by Sachs (2005, 2007), Kahl (2006), Stern (2007), and Lee (2009), among others.

---

[12]See, for example, Stein (2010) and the work of the International Crisis Group at http://www.crisisgroup.org/ (accessed November 15, 2012). There are obvious exceptions, such as concerns about the impact of nuclear proliferation in the Middle East or Northeast Asia, or Arab–Israeli tensions.

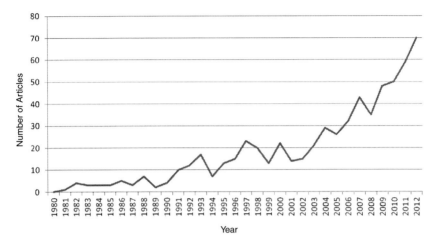

FIGURE 5-4 Growth in peer-reviewed literature on climate stress and armed political conflict, 1980–2012.*
NOTE: *2012 (through June) doubled to project annual rate.
SOURCE: Committee analysis using the SCOPUS database of peer-reviewed literature, July 2, 2012.
SOURCE NOTE: The search looked for some combination of climate-related and conflict-related words in the article's title, abstract, or list of key words. The climate-related terms were climate, rain, drought, flood, cyclone, hurricane, ENSO, El Niño, and water; the conflict-related terms were civil war, internal war, international war, transboundary conflict, transboundary violence, political conflict, political violence, armed conflict, armed clash, and genocide.

Some scholars have questioned the basis for linking climate change so directly to probabilities of increased conflict, internal as well as international, and, more broadly, to social and political instability. They question the conceptual and empirical bases of the arguments as well as the methodologies employed. The first major collection of peer-reviewed cautionary literature on the posited linkage between climate change and violence was a special 2007 issue of the cross-disciplinary journal *Political Geography* edited by Nordås and Gleditsch; it found no systematic empirical connections between climate change and conflict, although the editors noted the need for much further research (Nordås and Gleditsch, 2007). Salehyan (2008) followed that collection with a survey of the state of the literature whose title indicated its cautious conclusion: "From Climate Change to Conflict? No Consensus Yet."

In a study commissioned by the World Bank, Buhaug et al. (2010) came to a similarly cautious conclusion, noting that "numerous questions are unanswered regarding the proposed causal association between climate change and conflict" (p. 75). However, these researchers also offered a multi-step "synthesized causal model" to link climate change with conflict, which has the same sort of complex, contingent relationships as the framework we presented in Chapter 2. Their model proposes that "adverse climate change" could lead to increasing natural disasters, rising sea levels, and worsening resource scarcities, all three of which are posited to lead directly to increased or forced migration and then, both directly and indirectly, to "loss of economic activity, food insecurity, and reduction in livelihoods" (p. 82). The model also identifies such pre-existing conditions as poor governance, societal inequalities, and "bad neighbors" (countries characterized by ongoing violence) as well as population pressure exacerbated by migration, and it offers five "social effects of climate change [that] have been suggested as intermediating catalysts of organized violence": political instability, social fragmentation, economic instability, inappropriate response (possibly meaning inappropriate adaptation), and additional migration, all of which act in a feedback loop (p. 81).

The authors conclude by arguing that these five putative social effects of adverse climate change could lead to either increased opportunities to organize violence or increased motivation to instigate violence, with the end result being an increased risk of armed conflict. The authors repeatedly caution, however, that their model was intended for further research and testing purposes, and they emphasize that "whether adverse climatic changes result in *any of these social effects depends largely on the characteristics of the affected area*" (p. 81, our emphasis).

In a special section on human conflict in *Science* in May 2012, Scheffran et al. came to conclusions similar to those of Buhaug and his colleagues, in particular that "current debates over the relation between climate change and conflict originate in a lack of data, as well as the complexity of pathways connecting the two phenomena" (p. 869). They offer a different—but similarly complex—model with multiple potential causal linkages, and they provide a list of core research questions that need to be explored.

Another major set of papers in the peer-reviewed literature appeared in the February 2012 special issue of the *Journal of Peace Research* titled "Climate Change and Conflict." Guest editor N.P. Gleditsch introduced the issue by noting that while violence in general "is on the wane in human affairs, even if slowly and irregularly" (citing Goldstein, 2011, and Pinker, 2011), recently "pundits and politicians, along with a few scholars, have raised the specter that this . . . trend . . . might be reversed by environmental change generally and by climate change specifically" (Gleditsch, 2012:3). His overall assessment of the special issue's papers was that:

> [I]t seems fair to say that so far there is not yet much evidence for climate change as an important driver of conflict. In recent reviews of the literature, Bernauer, Böhmelt & Koubi (2012) and Gleditsch, Buhaug & Theisen

(2011) conclude that although environmental change may under certain circumstances increase the risk of violent conflict, the existing evidence indicates that this is not generally the case. (p. 7)

He adds, however, that

One of the lessons that the large-N community could learn from proponents of case studies is the emphasis on interaction effects. Homer-Dixon (1994) and Kahl (2006) do not argue that environmental change generally and climate change specifically have a major impact on conflict—the effect plays out in interaction with exogenous conflict-promoting factors (Buhaug, Gleditsch, and Theisen, 2008, 2010). Koubi et al. (2012) and Tir and Stinnett (2012) take a step in this direction in testing for interactions with institutions and regime type respectively. (p. 6)

Several of the articles in the special issue do offer more nuanced conclusions. In their modeling of range wars among pastoral groups in East Africa, for example, Butler and Gates (2012) are careful to note that they are really examining "weather change, particularly . . . drought." They conclude that conflict is actually more likely in situations of water abundance than in situations of water scarcity (in East Africa at least) and that the role of the state in defining and equitably administering rights to water is crucial to either outcome. In another of the special issue's articles, Hendrix and Salehyan (2012) report the results of an analysis that employed a new database of more than 6,000 instances of social conflict (including low-level conflict) in East Africa between 1997 and 2009 combined with rainfall variability measures for the same period. The authors find "a curvilinear relationship between rainfall and social conflict" and conclude that (consistent with Butler and Gates above) "armed conflict is more likely to break out in wetter years" (p. 46). They also argue that because so many African agricultural economies are "especially sensitive to rainfall shocks" and have "low adaptive capacity," climate change will have marked, but highly varying, effects on that continent.

Using more disaggregated conflict data from East Africa for the same time period (1997–2009), Raleigh and Kniveton (2012) explore rebel (antigovernment) versus communal violence and find that:

[A]nomalous rainfall conditions, irrespective of sign, are likely to enhance the probability of conflict. However, . . . the highest incidence of rebel conflict appears to occur in extreme dry rather than wet conditions . . . [but] incidences of communal violence appear to occur in extreme wet rather than dry conditions. (p. 62)

Then drawing from two case studies of pastoralist conflict (raiding, in particular) in northern Kenya, Adano et al. (2012) find that "more conflicts and killings take place in wet seasons of relative abundance, and less in dry season times of relative scarcity, when people reconcile their differences and cooperate" (p. 77). More relevant for possible generalizing, however, is how Adano et al. capture the importance of local coping mechanisms:

> During drought periods, pastoralists in northern Kenya deploy social institutions that mediate agency toward cooperation and guarantee access to resources (water) for all, thereby reducing violent conflict. Remoteness and inaccessibility . . . weaken government initiatives to provide adequate security, but local arrangements moderate conflicts when scarcity peaks. (p. 77)

The problem with this somewhat optimistic picture from Adano et al. (see also Solnit, 2010), is that it does not consider what would happen if climate change were to induce some combination of increased rainfall variability and immigration of affected peoples from other areas. This would result in a not inconsiderable challenge. How quickly and effectively, then, could the local social coping mechanisms adapt to the changing situation? Local coping mechanisms are usually borne of relatively static, or at least bounded, conditions and numbers of players with years of patterned interactions, but climate change may change both of those parameters.

Two other studies that looked at broader sets of data come to contradictory conclusions. In a study of the relationship between conflict and the ENSO cycle Hsiang et al. (2011) examined civil conflict data for the 1950–2004 period and found that "the probability of new civil conflicts arising throughout the tropics doubles during El Niño years relative to La Niña years" and that "ENSO may have had a role in 21 percent of all civil conflicts since 1950" (p. 438). This article was picked up by the mass media and attracted significant attention. Theisen et al. (2011), in an article in the journal *International Security,* use a "high-resolution gridded dataset of Africa from 1960 to 2004 that combines georeferenced and annualized precipitation data with new data on the point location of civil war onset and the location and political status of ethnic groups to test the links between drought and the start of civil conflict" (p. 81). They conclude, "The results presented in this article demonstrate that there is no direct, short-term relationship between drought and civil war onset, even within contexts presumed most conducive to violence" (p. 105). They also suggest, however, that

> future research needs to apply a broader understanding of political violence and armed conflict than is normally the case today. Given data limitations and a perception that major, state-based conflicts carry greater potential for political instability and state collapse than small-scale interethnic skirmishes, recent scholarship has focused almost exclusively on civil wars. This is reflected in the contemporary discourse on climate security, which is dominated by a state-centric approach. In contrast, nar-

ratives and news reports of conflict over diminishing resources frequently concern clashes between rivaling ethnic groups or between pastoralists and sedentary farmers. The conflicts in Assam in India, Darfur in Sudan, Kenya, Mali, and Mauritania, all central cases in the environmental security literature, were at least initially interethnic conflicts without explicit state involvement. Key questions in this regard are how environmental conditions and rapid environmental change affect intercommunal relations and local land use disputes, and what role the state plays in ending or fueling these conflicts. (p. 106)

Because climate change has the potential for an increase in the number or severity of various types of disasters caused by weather-related extreme events (cyclones, storm surges, floods, droughts, wildfires, etc.) or geographic shifts, or at least an expansion of their areas of incidence, there has been a renewed interest in the possible link between such events and interstate and intrastate violence.[13] Using a time series for 1966–1980, Drury and Olson (1998) provided the first quantitative attempt to test for a relationship between disasters and political instability and found "a direct and positive linkage between disaster severity and ensuing levels of political unrest" (p. 153). Ten years later Nel and Righarts (2008) analyzed a much larger number of cases (183 from the period 1950–2000) and found a positive and robust relationship between natural disasters of all types and both major (more than 1,000 killed) and minor (less than 1,000 killed) internal armed conflict occurring in the same year as the disaster as well as in the following year. Interestingly, when the analysis was limited to climate-type disasters only, there was only a correlation with major armed conflict, not minor. In a separate study that focused only on earthquakes, Brancati (2007) found a positive relationship between earthquakes and ensuing instances of political violence. The evidence indicates that climate events can contribute to social and political disruption in various ways, sometimes causing as much as a doubling of risks of adverse outcomes.

Recently, however, all of these findings and conclusions have been challenged by Omelicheva (2011), Bergholt and Lujala (2012), and Slettebak (2012), who question in different ways the previous studies' variable specifications and measurements, particularly their inadequate inclusion of control variables in their models. On the whole, their arguments contend that when closer attention is paid to variable specification and measurement and the models are made more complex with closer attention to such things

---

[13]A separate stream of literature not treated here on the role of disasters/catastrophes, including ENSO-related events, in what might be called civilizational collapses or macro-system changes would include Davis (2002), Diamond (2005), Nur and Burgess (2008), Fagan (2009), and Johnson (2011) among others.

as regime type, prior instability, and governance capabilities, the political instability effects of disasters tend to disappear.[14]

Given the relatively early stage of development of the research field and the strong policy interest in the topic, these sorts of debates can be expected to continue. A search for complex and contingent relationships should improve the conceptual basis for future research and intelligence analysis.

CONCLUSIONS AND RECOMMENDATIONS

Evidence from the social science literature supports the general argument that *climate change can contribute to social and political stresses that create security risks, but that these risks are not caused by climate change alone. They result from the conjunction of climatic conditions that generate potentially disruptive events with a variety of socioeconomic and political conditions*. The effects of climate on security in the coming decade are therefore likely to be indirect and contingent, operating through effects on systems that support human well-being (e.g., food, water, or health systems) or on specific events and circumstances (violent conflicts, disruptive migrations), and to depend on other social, economic, environmental, and political conditions in the affected places. This assessment is consistent with the conclusions about climate–security connections that appear in most of the major policy and government assessments.

The strength of the evidence about the linkages between climate events and outcomes of security interest varies substantially within and across issue areas. A number of the linkages are tenuous or not well understood; others seem relatively robust. Some examples of such linkages are:

- There is a statistically significant correlation between some forms of climate stress and the onset of some forms of armed internal conflict, but in general the causal pathways are not well understood.
- Climate change is altering the host range for several disease vectors with the potential to cause major epidemics and perhaps pandemics, given global patterns of trade and travel.
- Climate change is expected to cause changes in some of the basic and proximate conditions that can lead to increases in water insecurity, with the potential to affect food and health security.

---

[14]A second stream of literature not treated here involves the broader question of how publics evaluate the performance of their leaders in disasters, whether they are simply unthinkingly "responsive" (or, perhaps better, "reactive") or more thoughtfully "attentive" (evaluative). This stream, with an overwhelming U.S. focus, was stimulated by Achen and Bartels (2004) and includes Malhotra and Kuo (2008), Healy and Malhotra (2009), and Gasper and Reeves (2011).

The empirical knowledge base on the connections between extreme events of many types, including climate events, and political instability or violence also suggests some hypotheses that are worthy of examination in future research. For example, the available evidence is consistent with the idea that climate events affecting places of national security interest to the United States are likely to create the potential for significant violence, conflict, or breakdown dependent upon seven factors:

1. the nature, breadth, or concentration and depth of pre-existing social and political grievances and stresses;
2. the nature, breadth, or concentration and depth of the immediate impacts of the climate event;
3. the socioeconomic, geographic, racial, ethnic, and religious profiles of the most exposed groups or subpopulations as well as their susceptibilities and coping capacities;
4. the ability and willingness of the incumbent government and its internal and external supporters to devise, publicize, and implement effective, transparent, and equitable short-term emergency response and then longer-term recovery plans;
5. the extent to which emergent or established anti-government or anti-regime movements or groups are able to take strategic or tactical advantage of grievances or problems related to responses to the event;
6. the type, breadth, and depth of legitimacy and support for authorities, the government, the regime, and the nation-state; and
7. the coercive and repressive capacities of the government and its willingness and ability to engage in and carry out repression.

We reiterate that the available evidence indicates that the relationships are complex and uncertain between the kinds of climate events that can be expected to occur with greater frequency in the coming decade and the kinds of social or political outcomes that can become U.S. national security concerns. The picture is blurry in part because both the climatic and the political events of concern have been infrequent until now, making analysis of their relationships difficult. Available evidence on several of these connections, however, points to the same general finding we reported in Chapter 4 regarding the causes of social and political stresses, namely, that *the effects of climatic events on outcomes of security significance are contingent on a variety of specific social, political, economic, and environmental conditions in affected places.* Thus, even with a more extensive body of climate experience to draw upon, it is unlikely that simple, straightforward conclusions will be found that reliably link a climate event of a particular type with a particular kind of effect on conflict or on key aspects of social well-being.

In our judgment, it would be inappropriate to conclude from the evidence reviewed here that climate change will have no effects. In fact, the evidence indicates that climate events can contribute to social and political disruption in various ways. The appropriate conclusion is as follows:

Conclusion 5.1: *It is prudent to expect that over the course of a decade some climate events—including single events, conjunctions of events occurring simultaneously or in sequence in particular locations, and events affecting globally integrated systems that provide for human well-being—will produce consequences that exceed the capacity of the affected societies or global systems to manage and that have global security implications serious enough to compel international response. It is also prudent to expect that such consequences will become more common further in the future.*

Conclusion 5.2: *The links between climate events and security outcomes are complex, contingent, and not understood nearly well enough to allow for prediction. However, the key linkages, as with societal disruptions, seem prominently to involve (a) exposures to potentially disruptive events directly or through globally integrated systems affecting human well-being and (b) vulnerabilities (i.e., susceptibility to harm and the effectiveness of coping, response, and recovery efforts). In addition, security outcomes depend on the reactions of social and political systems to actual or perceived inadequacies of response.*

Available knowledge of climate–security connections that feature societal vulnerabilities, as reviewed in this and the previous chapters, indicates that security analysis needs to develop more nuanced understanding of the conditions—largely, social, political, and economic conditions—under which particular climate events are and are not likely to lead to particular kinds of social and political stresses and under which such events and responses to them are and are not likely to lead to significant security threats.

Recommendation 5.1: *The intelligence community should participate in a whole-of-government effort to inform choices about adapting to and reducing vulnerability to climate change. As part of this effort, the intelligence community and other interested agencies should support research to improve understanding of the conditions under which climate-related natural disasters and disruptions of critical systems of life support do or do not lead to important security-relevant outcomes such as political instability, violent conflict, humanitarian disasters, and disruptive migration.*

A major focus of this research effort should be on understanding the connections between harm suffered from climate events and political and social outcomes of security concern. These connections, which are arguably the most important aspects of climate change from a national security perspective, have received relatively little scientific attention until now. The disaster research community, which has been the locus of research on the political effects of climate events, has not been well connected to the climate research community. Nevertheless, the available research strongly suggests some plausible hypotheses to examine, such as the one above concerning seven factors that may link climate events to political conflict and instability. Efforts should be made to test such hypotheses systematically against historical data and, as climate change proceeds, against experience. There is also a need for fundamental research on some of the concepts that link harm to political outcomes.

Although there is extensive research on some of the factors influencing the vulnerability of populations to singular climatic events of various kinds, further investigation is needed to identify factors that influence vulnerability to sequences of events, such as repeated extreme precipitation events or linked physical and biological events driven by climate processes, and to events that occur in distant regions and disrupt food, energy, or strategic-product supply chains. There is also a need to develop real-time, local-scale metrics of key economic, social, and political components of vulnerability, as discussed further in Chapter 6 and Appendix E.

This research will need to use various methods and approaches. For example, given the complex and contingent relationships between climate events and such consequences as socioeconomic stress and political instability, *a systematic set of longitudinal case studies is needed* of the effects of climate events, using an explicit and common conceptual framework. These case studies need to cover at least five years post-impact and to include cases where an extreme event or several events produce no evidence of major socioeconomic or political stresses ("null" cases). The cases should cover all hazard types, with a special subset on climate-related hazard types. There is also a need for relatively *large-N quantitative studies* that focus on types and levels of disruptive events; mediating variables related to vulnerability, coping, and response that track multiple time periods; and ensuing internal political unrest, instability, or violence. There is also a need for cross-national, cross-cultural, and longitudinal *public opinion research* related to pre-event risk reduction and post-event coping, emergency response, and recovery in order to gain understanding of the factors affecting perceptions of adequacy of response.

We note that the needed knowledge tends to come from different communities of experts, which will need to communicate with each other but do not necessarily do so now. The recommended interagency process can help bring these communities of experts together, because they tend to associate with different groups of agencies.

6

Methods for Assessing National Security Threats

The preceding chapters, following the conceptual framework presented in Chapter 2, examined and evaluated evidence about the relationships and mechanisms that could link climate change and climate events over the next decade to outcomes of importance to U.S. national security. This chapter draws upon that analysis to address a core element of the project's statement of task: identifying "variables that should be monitored and ways that indicators of climate change, impacts, and vulnerabilities might be developed and made useful to the intelligence community in assessing climate-related threats to U.S. national security." Our premise is that the intelligence community needs a monitoring system capable of (a) supporting a continuing series of assessments of the likelihood and nature of security threats arising as a result of climate events in combination with other conditions, (b) informing timely preventive measures, and (c) supporting emergency reaction. An effective system must integrate information about climate, social stress, and response variables. It must be based on carefully considered priorities regarding the variables to be included, provide sufficiently high-resolution measures in space and time for critical locations and systems, and be actively and continuously managed and improved.

It is important to note that this study deals only with the opportunities provided by monitoring based on open-source materials. We recognize that the intelligence community also has access to classified information, but we do not have a basis for assessing how that relates to open-source materials.

We reemphasize that, as discussed in Chapter 1, there are important factors this study did not address that will also affect climate–security connections. In particular, countries adopt policies that may interact with climate events in ways that create security threats. For example, U.S. agricultural price interventions, as in use of corn for fuel, can have a direct impact on food prices in fragile societies that can be amplified by climate events that reduce agricultural production. Decisions to protect a country against climate events can also create or accelerate crises; for example, a unilateral decision by Turkey regarding the management of headwaters for the Tigris–Euphrates system could immediately generate crisis conditions. Another example is climate geo-engineering, which is already attracting considerable attention (Royal Society, 2009) and may become a source of conflict. In addition to the monitoring capacity we recommend here, the intelligence community will need the means to monitor, understand, and make forecasts concerning such developments.

WHAT SHOULD BE MONITORED, AND WHY

The analysis in the previous chapters, developed from our conceptual framework which was informed by an analysis of available knowledge, indicates that from a natural security perspective the climate events of most concern are those that would create the equivalent of a perfect storm: a country or region of importance to U.S. national security that experiences an extreme climate-related event or the effects of a climate-related shock to a global system that meets a critical need, that has significant human and economic assets in harm's way, where those assets are highly susceptible to harm, where local coping ability is static or decreasing, and where official response systems prove to be ineffective. In order to assess the likelihood that a country or region will experience a conjunction of these factors, all of these dimensions of security risk need to be monitored, that is, data about all these phenomena need to be repeatedly collected and examined.

Conclusion 6.1: *Monitoring to anticipate national security risks related to climate events should focus on five key types of phenomena:*

1. *Climate events and related biophysical environment phenomena;*
2. *The exposures of human populations and the systems that provide food, water, health, and other essentials to life and well-being;*
3. *The susceptibilities of people, assets, and resources to harm from climate events;*
4. *The ability to cope with, respond to, and recover from shocks; and*
5. *The potential for outcomes of inadequate coping, response, and recovery to rise to the level of concern for U.S. national security.*

In the domain of climate and biophysical environment variables, it is particularly important to monitor and estimate the *likelihood of potentially disruptive climate events* occurring in countries and regions of security importance to the United States or affecting global systems that meet critical needs in those places. Climate science provides considerable expertise for identifying, monitoring, and estimating the likelihood of single disruptive physical events occurring in particular places. Other kinds of science are needed in conjunction with climate science to define the monitoring needs for events that are more than just physical, such as climate-driven increases in food prices or outbreaks of infectious disease. These other types of science are also important for defining methods for monitoring and anticipating clusters and sequences of potentially disruptive events that might affect particular regions of interest and for considering the potential for climate events to generate shocks to integrated global systems of potential national security importance.

To monitor *exposures to potentially disruptive events* requires an understanding of where events are likely to occur as well as of who and what is or will be present in those places. Climate science can tell us what to monitor to foresee where potentially disruptive events may occur; social sciences that focus on population dynamics, economic development, and migration can tell us what to monitor to foresee what will be in harm's way. Several of the social, economic, and political conditions that contribute to exposure can be projected with some confidence based on available data; of those that cannot, many can be monitored at the country level and below.

Monitoring the degree and nature of *susceptibility to harm from climate events* should focus especially on places and systems of security concern. It needs to consider different susceptibilities to different kinds of events as well as differences among populations separated by place or differentiated by class, race, ethnicity, religion, or other social cleavages.

Monitoring *the ability to cope with, respond to, and recover from shocks* requires measures or assessments of limitations in the capacity of affected people, communities, or sectors to cope on an informal basis as well as limitations to the ability or willingness of governments and other formal assistance organizations to respond after an event occurs. It also requires measures or assessments of the likelihood that responses to disruptive events, particularly by responsible governing authorities, will be (or be perceived to be) inadequate. Past performance in natural disasters may provide useful indicators for most of these factors; indicators of corruption or favoritism in the delivery of public services are particularly relevant for the last. Social science offers a variety of tools and methods for monitoring aspects of susceptibility, coping, response, and recovery. Normal techniques of intelligence analysis are also useful for assessing some of these components of vulnerability, such as the willingness of governments to respond on behalf of only particular segments of their populations in the event of need.

Monitoring the *potential that inadequate responses will rise to the level of concern for U.S. national security* entails estimating and assessing the ways security conditions in countries and regions of interest could be affected by climate events. A major potential link involves a combination of susceptibility and inadequate response leading to a humanitarian emergency, violence, or political instability. Climate events that disrupt the lives of affected populations are more likely to lead to larger upheavals when the events are serious, when governments underperform expectations in responding, when there is pre-existing dissatisfaction with the government, and when there are organized opposition groups positioned to use dissatisfactions as an opportunity to mobilize confrontations with authorities. The monitoring of many of these security conditions is a standard intelligence function and is related to the monitoring of state fragility.

Monitoring of these five types of phenomena would provide valuable input for national security analysis. Given that security threats arise from combinations of all of these, indicators and monitoring systems should be developed to follow them at various levels from local to national. Monitoring will also need to take into account the fact that some of the above types of phenomena are specific to certain kinds of climate events (e.g., flooding), while others, such as the capacity of emergency-response organizations, have an effect on consequences for many different types of hazards. As a rough generalization, exposures tend to be hazard-specific (e.g., some populations are exposed to coastal flooding and storm surge, others to inland drought), which implies that the monitoring for exposures should be differentiated by hazard type. Coping and response factors (e.g., funding and organizational effectiveness of disaster-response agencies) tend to be much less hazard-specific. Susceptibilities can be either hazard-specific or general. For example, the health status of a population provides an indicator of susceptibility to a variety of stresses (e.g., diseases and food shortages), whereas some attributes of infrastructure reduce susceptibility to only single hazards (e.g., to floods but not wildfires).

It is important to note that although most of the phenomena of all the types we have highlighted normally change on time-scales of months, years, or decades, potentially disruptive climate events often give far less warning. Monitoring of the slower-moving factors makes it possible to use a scenario approach for considering the consequences of rapidly appearing climate events. In this approach, analysts posit the occurrence of a particular potentially disruptive climate event the risk of which is high or increasing and consider how a country, region, or system of interest would likely respond, given what is known or expected from monitoring and assessment of the state of other environmental conditions, exposures, vulnerabilities, and likely responses to inadequacies of coping, response, and recovery. Monitoring and assessing these slower-moving variables will enable analysts to apply a sort of stress test to target countries, regions, or systems in order to anticipate the social, political, and security consequences that could arise if these events affect them. We return to this idea later in the chapter.

CHALLENGES OF MONITORING

Each of the five types of phenomena highlighted above encompasses many specific elements or variables that could be monitored. Appendix E presents and discusses a wide range of these. A major challenge of monitoring lies in setting priorities: determining which data and kinds of data are most needed. This task involves determining which of the many possible measures of a factor such as exposure to coastal flooding, susceptibility to malaria, or effectiveness of disaster response will be most reliable, valid, and useful as part of a larger monitoring system for assessing security threats.

Setting Data Priorities

Because of the multifaceted nature of the phenomena that might connect climate events to national security concerns and because of the complexity of these connections, setting priorities for monitoring is a significant challenge. Substantial ongoing monitoring activities may prove useful for measuring aspects of the key phenomena. Many are already being carried out by the intelligence community and other parts of the U.S. government as well as by various international organizations. However, much of this work has been organized for other purposes. In the area of climate change, the activities include efforts to forecast climate events and estimate the vulnerabilities to climate change of various aspects of human well-being. Outside of the climate change community, monitoring of environmental conditions and changes is carried out largely by environment agencies, and monitoring relevant to exposure and susceptibility is carried out largely by departments and organizations focused on development or disaster assistance. A number of these monitoring efforts are described in Appendix E.

There is strong reason to try to identify a small number of reliable and valid indicators to cover a great variety of phenomena. Some potentially useful indicators are already in use or in development in the U.S. intelligence community and elsewhere, and in many cases they are available in the open literature. It is important to emphasize, however, that the basis for constructing such indicators is quite uneven across variable types and across parts of the world. For example, it is possible to develop a map of the western United States with fairly sharp spatial resolution that indicates the risk for forest fires and related events as a function of projected increases in average temperature (see Figure 3-2), but much less confidence can be accorded to similar maps in parts of the world where there are not such well-developed databases on past fires and on temperatures.

Less well established is the practice of developing indicators for such things as the coping capacity of communities or the ability of governments to mobilize disaster response. For many of these factors, however, sufficient knowledge exists to identify some of the measures that might constitute an indicator and therefore to begin building and testing indicators. For example, a recent analysis of the determinants of political instability features the effects of political institutions and policies for allocating resources (Goldstone et al., 2010). That analysis identified 141 episodes of political instability that occurred between 1955 and 2003 and demonstrated that 80 percent of them could have been predicted two years in advance using an indicator that combined measures of regime type (degree of democracy), infant mortality, incidence of conflict in neighboring countries, and internal favoritism. The greatest likelihood of instability was associated with political regimes intermediate between democracy and autarchy that practiced discriminatory policies.

Despite progress in developing indicators of several of the key phenomena and the connections among them, it is premature to settle on a small number of variables to monitor that will be sufficient to meet the needs of analysis. Research, including experiments and pilot studies, will be required to determine which measures can serve as proxies for which others and thus to develop an efficient and effective monitoring system. In many instances, existing knowledge does not yet support reliance only on quantitative indicators.

Progress can be made by focusing on each class of phenomena separately. Appendix E illustrates the current state of thinking about data needs and discusses examples of current monitoring efforts for climate and biophysical variables, exposures, susceptibility to harm, and coping, response, and recovery. It shows that across these types of phenomena there are substantial differences in the level of consensus within the relevant communities of experts about which are the key variables from which a small and useful set of indicators could be developed.

An example of a high level of expert consensus, and one approach to consensus-building is the work of the climate science community to identify "essential climate variables" (ECVs). In 1998 the Intergovernmental Panel on Climate Change and the United Nations Framework Convention on Climate Change (UNFCCC) established specific requirements for systematic climate observations and a sustained observing system. The Global Climate Observing System, sponsored by the World Meteorological Organization; the United Nations Educational, Scientific and Cultural Organization; the United Nations Environmental Program; and the International Council for Science, is charged with advising the community on global climate observations and overseeing implementation based on UNFCCC standards. In 2010 the organization developed a list of 50 ECVs that are possible to monitor globally and whose observation could yield significant progress toward meeting the UNFCCC requirements (Global Climate Observing System, 2010; see Figure 6-1). As discussed further in Appendix E, a number of efforts are under way to use these variables to create a more limited set of indicators that could be more relevant to policy making. This indicates a good level of progress in priority setting. However, to our knowledge there has as yet been no serious effort at priority setting among climate measurements, or among environmental measurements more broadly, for the purpose of informing security analysis.

In other domains in which monitoring is needed, there is variation in the degree to which the relevant scientific communities have defined priorities for monitoring and in the degree to which the monitoring systems are relevant to the climate–security nexus. Two other examples of active monitoring efforts that reflect substantial consensus but also suggest the degree to which such systems learn and adapt by gathering and assessing data come from food security and public health.

An example of an active monitoring effort where experience over time has led to a new understanding of key variables and relationships is the Famine Early Warning Systems Network (FEWS NET). Created in 1985 and funded by the U.S. Agency for International Development, the project is a collaboration among national, regional, and international partners, including expert field personnel on the ground that monitor and analyze relevant data and information in terms of its impacts on livelihoods and markets to identify potential threats to food security. A range of products provide alerts, monthly status reports, outlooks, and in-depth studies based on FEWS NET's on-the-ground coverage of 23 countries, mostly in Africa, but including Central America, Haiti, and Afghanistan. Less extensive coverage is provided for 15 more countries through partner-based monitoring. FEWS NET, which relies on a combination of physical and socioeconomic indicators to estimate and predict the degree of, and changes in, the food security conditions of vulnerable countries, was one of the earliest users of satellite imagery to monitor rainfall and crop conditions in the developing world. It now also looks at longer-term global climate variability to help assess future threats to food security and inform priorities for climate adaptation activities. One result of its ground-breaking use of food prices and "commodity market networks, market integration, the geographic and economic distribution of food commodities, and cross border trade" (Famine Early Warning Systems Network, 2008:1) was the development of new ways of seeing the nature of socioeconomic threats to food security in some countries—significantly different patterns of food pricing within countries, strikingly divergent global food price influences in different re-

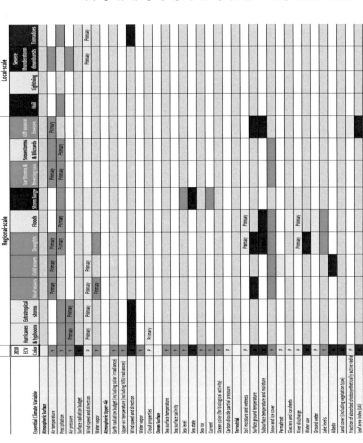

FIGURE 6-1 Relationship of extreme phenomena to essential climate variables (ECVs) for monitoring. Both the phenomena and the ECVs are color coded to describe the adequacy of the current monitoring systems for capturing trends on climate time-scales. Green indicates global coverage with a sufficient period of record, data quality, and metadata to enable meaningful monitoring of temporal changes. Yellow indicates an insufficiency in one of those three factors. Red indicates insufficiency in more than one of the factors. In the left column, Y indicates "Yes, it is adequate," N indicates "No, it is not adequate," and P indicates "Partial adequacy." The word "primary" in the colored ECV block indicates that the ECV is of primary importance to monitoring changes in the extreme event phenomenon.
SOURCE: Trenberth et al. (2012). Courtesy of James McMahon.

gions, and local prices that track closely with global prices in some places but not in others.

Monitoring and evaluation efforts for public health began more than 150 years ago. Emerging infectious diseases are receiving increasing attention because there is a long history of such diseases surprising societies and causing high morbidity and mortality, sometimes accompanied by social disruption (Lindgren et al., 2012). In response to a presidential directive in 1996, the U.S. Department of Defense established a Global Emerging Infections Surveillance and Response System, with the mission to monitor newly emerging and re-emerging infectious diseases among U.S. service members and dependent populations (Clinton, 1996). The U.S. Armed Forces Health Surveillance Center, Division of Global Emerging Infections Surveillance and Response System Operations coordinates a multidisciplinary program to support the International Health Regulations (IHR).[1] The goal is to link datasets and information into a predictive surveillance program that generates advisories and alerts on emerging infectious disease outbreaks (Witt et al., 2011). Datasets and information are derived from eco-climatic remote sensing activities; ecologic niche modeling; and arthropod vector, animal–disease host/reservoir, and human disease surveillance for febrile illnesses. The program includes 39 funded partners working in 92 countries (Russell et al., 2011). Other organizations monitoring emerging infectious diseases include the U.S. Centers for Disease Control and Prevention, the European Centre for Disease Control, the WHO Global Alert and Response network (including the Global Outbreak Alert and Response Network), and the PROMED reporting network at the International Society for Infectious Diseases. These activities could be enhanced to consider how climate variability and change could alter the risks of outbreaks in geographic regions of interest.

Recommendation 6.1: *The intelligence community should participate in a whole-of-government effort to inform choices about adapting to and reducing vulnerability to climate change. One of the objectives of this effort should be to build the scientific basis for indicators in this domain.*

This effort would support activities by the research communities involved in assessing exposures and vulnerabilities to environmental change to identify a relatively small number of key variables relevant to the social and political consequences of climate events. The effort of the climate science community

---

[1] The IHR, adopted in 2007, establish a mandatory reporting system by all 194 members of the World Health Organization for events that may constitute a public health emergency of international concern.

to identify a small number of "essential climate variables" suggests the kind of process that could be used. A recent effort by the National Research Council (2010b) took an initial step in this direction. The recommended effort might identify sets of variables to monitor that could become elements of indicators of exposures to such events, susceptibilities to harm, and of the likely effectiveness of coping, response, and recovery efforts at the levels of communities, countries, and systems that support critical human needs. It would also support research to develop and validate indicators of key phenomena linking climate and security, as has been done with research on the phenomena of political instability.

The Role of Quantitative Data

Quantitative indicators that combine multiple datasets are often highly useful for giving decision makers a broad picture of a phenomenon of concern. For example, composite indicators are sometimes created to summarize knowledge about various phenomena of interest, such as drought, susceptibility to damage from flooding, health status of a population, emergency coping capacity, or political instability. Developing such indicators for the full range of security threats related to climate change presents a major challenge for several reasons.

Integrating Data Types

Observations relevant to climate-security linkages may come from a great diversity of sources: scientific instruments such as space satellites or ground-based sensors, censuses or surveys conducted by governments or nongovernmental organizations, scanning of communications on mass media or the Internet, and on-site expert observation of qualitative aspects of social and political systems, among others. Some of these data sources are quantitative and others qualitative. Of the quantitative indicators, some are already well calibrated and validated, others much less so. For example, data on some socioeconomic factors, such as demographics and gross domestic product, are routinely collected by well-developed methods. Other data, such as on the ability of a region or country to cope with an extreme event of a particular magnitude or on the condition of resource stockpiles, are often collected through surveys. Because the design, conduct, analysis, and archiving of survey data can be time-consuming and costly, it would be prudent to determine in advance what types of data are likely to be needed and how often the survey should be repeated for critical information, taking into consideration how data might be relevant. This not only would be good planning but also could offer opportunities to identify surveys already being conducted to which additional components could be added.

Monitoring systems will require the integration of quantitative indicators of both environmental and social phenomena, qualitative data on social systems, and information gathered by traditional security and intelligence analytic methods. Integration will be necessary in part because different methods are necessary to gather different kinds of information. Some important kinds of information, such as how governments are likely to respond to disasters, are difficult to collect because many governments will be responding without elaborate advance planning or training. Also, different methods of validation are appropriate for different kinds of data and information (see, e.g., King et al., 1994; George and Bennett, 2005; Brady and Collier, 2010). All methods should be used to gain insight. Where the same kind of information can be obtained by multiple methods, this situation creates an opportunity to use each method as a check against the others and thus increase overall confidence in assessments. We note that it may be possible to advance the objectives of both analysis and risk reduction if information gathering is done through open dialogue, sometimes with the assistance of the U.S. government.

Developing the needed broad monitoring system for climate-related security threats will also require integration of climate science, various branches of social science, and security analysis. The integration of the social science of natural disasters and disaster response with other forms of analysis will be particularly important for assessing the security consequences of climate change because many disruptive climate events will be perceived and responded to as natural disasters.

Judgment is involved in creating indicators, even when they are built on highly reliable observations. Expert assessment is needed of the accuracy and validity of indicators and of whether other relevant information should be taken into consideration. For example, weather forecasts are only partially based on the vast amounts of data analyzed by highly sophisticated computer models; skilled meteorologists modify the forecasts based on an understanding of the complex weather systems that extends beyond what can be coded into a model and of the performance of a particular computer model for a weather variable in a specific region.

Coverage and Resolution

Because of the various purposes for which data collection has been organized, existing data may or may not have the coverage or offer the degree of spatial or temporal resolution needed to track and analyze the key variable sets in ways needed for security analysis. In many cases, data need to be collected at higher resolution than in the past in order to support improved analytic assessment. High-resolution monitoring will be especially important for highly significant and highly vulnerable locations.

The appropriate level of spatial and temporal resolution for indicators varies, however, with the substantive domain. Some indicators, such as of the capacity of national governments to provide response and recovery support after a disruptive event, are most appropriately measured at the national level. Others, such as the risk of coastal flooding, may need much finer resolution, especially in areas of dense population. It may be necessary to develop indicators of community coping capacity separately for communities defined by geography and for communities defined by business relationships or cultural similarities. Similarly, risk indicators may require more frequent updating for some climate events than for others. The need for temporal resolution is probably greatest during the development of a slow-moving climate event, such as a drought, and in the immediate aftermath of a disruptive climate event.

For many existing and potential indicators the required spatial and temporal resolution is finer than what is currently available. In setting priorities for indicator development and improvement, the intelligence community should take into account the gaps between the existing and the desired resolution and should invest in improved resolution for those indicators judged to be the most needed and the most useful in places of concern. When considering how to invest in the development of indicators focused on a particular country or region, it will be worth considering the extent to which an indicator could be applied elsewhere. It should also be kept in mind that existing datasets may provide—or may be analyzed to provide—useful information. Over time, determining the needed data coverage and resolution tends to proceed from an initial assessment of the main data needs and to evolve as the monitoring system is used.

Validating Indicators

Validation involves determining the extent to which an indicator actually measures the phenomenon it purports to measure—a task that can be extremely challenging. Validation is a long-term process, especially for measures associated with the likelihood and consequences of disruptive events, which are almost by definition uncommon. A measure of the likelihood of a climate event that occurs only once in several decades may require a century or more for full validation, so validation efforts might involve using one indicator to validate another. For example, a projection of drought risk from climate models might be tested against a drought severity indicator and validated even in the absence of extreme drought events. A measure of response capacity may have to wait for validation until a disruptive event occurs, but if it holds up in two or three disruptive events, confidence in its validity should increase. As already noted, the validation of indicators for many of the factors linking climate change and security will not be a mechanical exercise but rather will likely require considerable judgment for some time to come.²

It is important to develop and validate monitoring systems now in order to have baseline data for future studies of climate events, for the associated exposures and elements of vulnerability, and for social and political stress analyses. Open data sharing, information regarding source codes, and transparency in the analyses are also essential elements of this process. Validation is particularly an issue with emerging monitoring technologies, such as those involving sophisticated data-mining algorithms (e.g., of Internet postings) or remote observations that are overlaid on geographic information systems. Such techniques may produce outputs that catch the eye and are very impressive on first glance, but they are sometimes closely held by their developers and difficult to validate, especially if they involve infrequent events. Indicators and monitoring results should be interpreted with caution until these techniques develop a record of validation.

Improvement of Indicators and Analytic Techniques

Each country presents its own unique mix of exposures, vulnerabilities, socioeconomic conditions, and so forth, and different kinds of climate events have different associated patterns of exposure and vulnerability. There is no formula to tell intelligence analysts where to focus monitoring efforts and which variables are most important to monitor.

As climate change proceeds and its human implications continue to be experienced, a clearer understanding is likely to emerge of the mechanisms linking climate events to security concerns and, therefore, of the most important things to monitor. The needed monitoring will be a major undertaking over an extended period, which should avoid duplication of

---

²Several persistent and sometimes specialized statistical issues will also need serious attention as part of the development of information that will be useful to the intelligence community. One of particular importance for this study, already discussed in Chapter 3, is dealing with rare events, which introduce a variety of subtle, although well studied, statistical problems. Another is the widely recognized but difficult problem of making estimates from data of widely differing quality when that quality varies systematically rather than randomly, for example, across regions, types of government, and level of economic development. A third is the sophisticated use of statistical controls—both classical control variables and the newer methods for matching cases—to improve estimation. Closely related to this is the application of quasi-experimental design methods and identifying cases where these are appropriate.

More broadly, as the Political Instability Task Force forecasting tournament indicated, there is a need to identify the relative merits of the competing data analysis approaches currently prevalent: frequentist (relying on significance testing); Bayesian (using probability distributions); and machine learning (pattern recognition, broadly defined). Related to this issue is the question of applying an assortment of new, computationally intensive ensemble methods that integrate the results of multiple models, such as Bayesian model averaging.

effort and strive for maximum amounts of synergy and complementarity. Existing databases should be identified and consulted before new ones are developed; feedback should be provided to clarify what indices are most and least useful; and all datasets should be as transparent and accessible as possible so that analysts working on all dimensions of climate change can rely on the work of colleagues and use it to address their own questions of greatest interest—in this case, questions about security issues.

As important as data and monitoring are for assessing the effects of potential climate events on national security concerns, data in the absence of effective analytical techniques to process them and produce useful information are of little help. In some situations, more data could actually make the situation worse in terms of producing useful forecasts. Some of the major challenges are associated with two questions.

What analytical techniques are most appropriate for the growing number of "big data" approaches? Advances in computing power and new developments in data mining have the potential to allow the intelligence community to gather more data about more places more quickly than could be imagined even a few years ago. Perhaps the best example of this challenge is the explosive growth of different forms of social media via mobile technology, which is yielding, essentially in real time, potential new forms of data about political and social trends. The development of sophisticated algorithms is making it possible to perform machine-based analysis of data from social media and other kinds of events data instead of relying on the traditional laborious coding by individuals; meanwhile, advances in translation software enable the monitoring of many more sources of information from all forms of media.

Applying these new technologies will require successfully addressing issues of data quality and reliability, interoperability across databases and different forms of data, risks of false positives, and the sorting out of meaningful signals from large amounts of noise. At present it has to be acknowledged that the capacity to acquire data in many cases exceeds the analytical capacity to translate the data into useful information that can improve understanding of trends in key countries or regions. For example, the development of analytical—as distinct from visualization—methods for social network and geolocated data is in its infancy, compared with techniques for the analysis of older forms of data, such as econometric or survey.

How much data are really needed? Contrary to the assumption that more is better, with statistical analysis, additional variables may simply duplicate what other data already provide or in some cases actually decrease the accuracy of a model because of poor data quality or other statistical quirks. The pursuit of parsimony, or the "reduction of dimensionality," is important in numerical analysis. The State Failure Project, the predecessor to the current Political Instability Task Force (PITF), began working with a set of some 700 potential variables and over time reduced its final model to 4. (See Box 5-1; a list of many of the variables examined may be found in Goldstone et al., 2010.) Another example of reducing dimensionality can be found in the work of Cutter and colleagues (Cutter et al., 2003; Cutter and Finch, 2008) to develop an indicator of social vulnerability to environmental hazards. Using spatial and county-level data in the United States on 30 variables relevant to vulnerability (e.g., wealth, employment structure, demographic composition, and other factors known to influence a community's susceptibility and response capacity), they determined that the variables could be represented by 7 underlying principal components, which were summed to create a single value for each county of a social vulnerability index (SOVI; Hazards and Vulnerability Research Institute, 2012).

Conclusion 6.2: Developing an adequate system for monitoring the conditions that can link climate events to national security concerns will require maintaining critical existing observational systems, programs, and databases; the collection of new data; the analysis of new and existing data; and the improvement of analytic systems, leading to a better understanding of the linkages over time and to improved indicators of key variables where quantitative indicators are appropriate and feasible to produce. It will typically require finer-grained data than are currently available. It will also require improved techniques for integrating quantitative and qualitative information.

We emphasize that improved understanding and monitoring of the various elements of climate vulnerability—a key link between climate events and security concerns—is an objective that the intelligence community shares with the U.S. Global Change Research Program (USGCRP) and many other institutions at federal, state, local, and international levels.

The intelligence community cannot address these challenges alone. Addressing many of the new and enduring methodological problems is largely the province of the academic research community. The intelligence community needs to draw on this knowledge, as efforts like the PITF are doing, to address the interactions of climate events with traditional intelligence community concerns.

A STRATEGY FOR MONITORING

The United States, like other countries, lacks a national strategy for sustained, long-term observations for the purpose of informing analysis of relationships between environmental changes, including climate change, and national security. Multiple environmental monitoring activities and programs exist, organized by both public and private actors; they have diverse purposes and are focused on conditions and processes that range from local to global. Only a few of them are organized to inform security analysis, however, and it is difficult to know how useful the others might be for that purpose. Efforts to develop environmental observation priorities for security analysis should focus on identifying a small number of composite indices designed for specified purposes of analysis or early warning. The same can be said for social, economic, and political observations: Multiple monitoring programs exist, with diverse purposes, and only a few of these are organized to inform security analysis. Organized efforts at indicator development for climate–security analyses remain works in progress. Yet systematic efforts are needed. Progress will require additional work, which should be conducted through collaborations involving climate scientists, environmental scientists, social scientists, and security analysts.

The intelligence community should adopt a risk-based strategy for setting its monitoring priorities. Such a strategy seeks to prioritize the measurement and assessment of the most significant expected security risks that may arise from conjunctions of potentially disruptive climate events; exposures; susceptibilities; limitations of coping, response, and recovery; and the reactions to revealed limitations. A strategy that is risk-based considers the product of the likelihoods of events and the magnitude of their consequences. However, because the likelihoods of key events—and even in some instances the nature of the events—are not well known, monitoring under a risk-based strategy is not an exact science and must be expected to evolve as research and monitoring activities improve understanding of which conditions are most important to monitor and provide increasingly valid estimates of the probabilities and consequences of key events.

Threat Monitoring as a Long-Term Research Activity

Developing a monitoring system for climate-related security threats is a long-term enterprise. As noted above, a considerable amount of effort is already being devoted to monitoring climate events and trends; some aspects of food, water, and health security; risks of natural disasters related to climate change; and certain elements of disaster response capacity by a variety of governmental, nongovernmental, and international organizations for various purposes. Such existing monitoring systems, both open-source and commercial, should be periodically scanned for potential usefulness, but with critical attention paid to indicator selection, data reliability and validity, and cross-case and cross-national comparability.

As we have also noted, the connections between climate events and national security concerns are complex and contingent, with many plausible combinations of climatic events with social, economic, and political conditions that might create risks to U.S. national security. These risks are unlikely to be foreseen by looking only at climate trends and projections or by looking only at political and social trends and projections. *To anticipate the risks, analysis needs to integrate three kinds and sources of knowledge: (1) knowledge of political and socioeconomic conditions in countries of interest, (2) knowledge from climate science about the potential exposure of these countries to climate events, and (3) knowledge from social science about the susceptibility of these countries to be harmed by those events and the likelihood of effective coping, response, and recovery at local to national levels.* These sources of knowledge come from different communities of experts, which will need to communicate with each other. Making this happen will take time and continued effort.

Indicators based on monitoring efforts can be used even while research and development on them is in an early phase if they are interpreted cautiously as one source of insight among many, including qualitative insight derived from on-the-ground information and experience. Open-source monitoring efforts can help reduce the risks of climate change by helping national and international decision makers anticipate potentially disruptive events and reduce vulnerabilities. Monitoring efforts by the U.S. intelligence community may also have such broader benefits.

Efforts to develop quantitative indicators need to be improved over time to maximize their usefulness for security analysis, and achieving this goal will require a long-term effort with a significant research component. As such indicators are developed and validated, it will become appropriate to assign more weight to the information and predictions they provide. The intelligence community should consider the development of the needed indicators to be a long-range research activity.

A research investment in indicator development is likely to increase in value over time, both because monitoring systems are likely to improve through continued efforts and because potentially disruptive climate events are expected to increase in frequency and intensity in years and decades to come. *It is therefore important to begin now to build and test the capability to monitor and anticipate climate-driven security threats.* The potential for disruptive events, the elements of vulnerability, and security conditions will all need continued monitoring because they are all changing and can affect each other. For example, responses to recent climate events or other disasters can affect both the future capacity to respond and security-related conditions, such as public support for governments. The research effort needs to integrate monitoring across variable types and methods and should focus on validating indicators, monitoring the appropriate spatial and temporal resolution, and improving analytical techniques, particularly to make effective use of rapidly increasing volumes of data.

The Need for a Whole-Government Approach

Recommendation 6.2: *The U.S. government should begin immediately to develop a systematic and enduring whole-of-government strategy for monitoring threats connected to climate change. This strategy should be developed along with the development of priorities and support for research as recommended in Chapters 3, 4, and 5.*

The monitoring should include climate phenomena, exposures and vulnerabilities, and factors that might link aspects of climate and vulnerability to important security outcomes, and it should be applicable to climate issues globally. It should also include making and periodically updating priority judgments about when and where high-resolution monitoring is needed.

The recommendation for a whole-of-government approach is consistent with the recommendations of the Defense Science Board (2011) and the strong convergence of the climate change monitoring objectives of the intelligence community as discussed here and those of the USGCRP. As noted in previous chapters, these interagency enterprises have many common needs for monitoring and for the fundamental science that informs monitoring choices, but their efforts are not integrated. As the recent National Research Council (2012a) review of the USGCRP strategic plan noted, "An effective global change research enterprise requires an integrated observational system that connects observations of the physical environment with a wide variety of social and ecological observations. Such a system is a crucial foundation for identifying and tracking global changes; for evaluating the drivers, vulnerabilities, and responses to such changes; and for identifying opportunities to increase the resilience of both human and natural systems" (National Research Council, 2012a:39). Monitoring for the purposes of the intelligence community has the same requirements, although information will be used differently because of the need to focus on threats outside the United States. It makes sense for these different interagency communities to collaborate on the scientific analysis required to design the needed monitoring and assessment systems and, as appropriate, on the development and use of these systems.

Organized international collaboration with potentially affected societies and governments and open sharing of data will be important aspects of developing the needed monitoring systems. A monitoring system capable of anticipating and detecting severe instances of climate-induced social and political stresses in many countries would be of great value not only to the U.S. national security community, but also to the affected countries themselves to guide anticipatory adaptations as well as to international humanitarian assistance agencies and foreign donors for preparing their response capacities and to security analysts in countries other than the United States and the affected countries as they consider the security implications of climate events. Such a monitoring system with open sharing of data would thus provide a global public good. The U.S. government would also benefit from data-gathering efforts in and by other countries.

Open, international scientific collaborations are also desirable on scientific grounds. The development of compatible concepts, databases, and indicators across countries helps speed scientific progress and improves the ability to learn from experiences in other countries.

International collaboration is likely to be necessary to achieve acceptance of higher-resolution monitoring at critically vulnerable locations, particularly if that monitoring requires an on-site component. Such a system would inherently include elements that could be seen as intrusive in the countries being monitored. Thus its global acceptability would depend on justifying its purposes and legitimizing its rules. In particular, such a system would have to be credibly directed to broad common interests rather than intended to provide some competitive national advantage that might be perceived as hostile. If the capacity of a society to manage internal stress is to be subjected to detailed scrutiny, the motivating purposes must be accepted as constructive, access to the data must be equitable, and the benefits derived must be mutual. Given the historical legacy of security concerns, those conditions will not be easy to achieve, but they will certainly be essential.

As a practical matter these conditions would have to be established though a process of evolution as the details of monitoring arrangements are worked out. A mature system would almost certainly have to be achieved in a series of incremental steps. Nonetheless, transparency would be a central principle from the outset. To the maximum extent possible, both the methods used and the data resulting from a monitoring system must be open to global scrutiny as the best and, ultimately, the only way to establish legitimacy and to assure accuracy. That does not mean that access would be completely unrestricted. It means rather that the rules of access would be based on criteria that are broadly accepted at the outset, universally accepted in a mature system, and subject to collective reconsideration over time.

Of course, U.S. government agencies will continue to need to gather some kinds of information that will not be openly shared, and there will be questions about which data- and information-gathering methods can and should be openly shared. There will also be suspicion of the involvement of U.S. intelligence agencies in international information-gathering efforts related to security. Such issues will need to be addressed in ways that we have not had the opportunity to consider in this study. Nevertheless, the benefits of open, international data development and sharing should be taken seriously as work on monitoring systems proceeds.

AN APPROACH TO ANTICIPATING RISKS

Recommendation 6.3: *The intelligence community should establish a system of periodic "stress testing" for countries, regions, and critical global systems regarding their ability to manage potentially disruptive climate events of concern. Stress tests would focus on potentially disruptive conjunctions of climate events and socioeconomic and political conditions.*

The intelligence community presumably already uses an analogous process to consider the ability of foreign governments and societies to withstand various kinds of social and political stresses. This recommendation calls on the community to incorporate climate risks and the associated exposures and vulnerabilities into such exercises. The concept of a climate stress test provides a framework for integrating climate and social variables more systematically and consistently within national security analysis.

A stress test is an exercise to assess the likely effects on particular countries, populations, or systems of potentially disruptive climate events to which they have some likelihood of exposure in the coming decade. The recommended stress tests would involve analyzing the likely effects of an event at some projected time of occurrence in terms of key variables affecting susceptibility, coping, response, and recovery or the failure thereof, and the likely responses within regions or countries of interest in the event that these actions are perceived to be inadequate. The tests would draw on knowledge about the potential events and each of the other types of phenomena and would provide a major way of making knowledge about climate events, exposures, and vulnerabilities operational in security analysis.

Stress tests should consider two kinds of climate events of potential security concern: those that climate scientists can say with some confidence are increasingly likely to occur or become more severe, and those that seem increasingly likely to occur based on a fundamental understanding of climate dynamics but about which available evidence is not yet sufficient for climate scientists to attach confidence to such projections. Stress tests might also be triggered by assessments indicating that event likelihood, exposure, or susceptibility is increasing or that the capacity to respond adequately to certain kinds of climate events is declining in a region or country of concern.

The results of stress tests would inform decision makers about places that are at risk of becoming security concerns as a result of climate events and could be used by the U.S. government or international aid agencies to target high-risk places for efforts to reduce susceptibilities or to improve coping, response, and recovery capacities. The stress testing process would also help examine and refine hypotheses, such as those presented in Chapter

5, about the characteristics of climate events and of the affected societies that determine whether or not potentially disruptive climate events turn into security threats. Over time an accumulation of data on potentially disruptive events and their social, political, and security consequences will improve understanding and feed back into improved monitoring processes and improved skill in stress testing.

Countries, regions, and systems of particular security interest should be prime targets for periodic stress testing. Given the joint criteria of significant potential for climate change impacts and importance to U.S. national security, it is likely that no more than 12 to 15 countries will need to be monitored and subjected to periodic stress tests over the next decade, many of which are likely to be in critical, and often shared, watershed areas in South Asia, the Middle East, and Africa. If the criteria for importance to the United States are expanded to include foreign policy and humanitarian concerns, the number of countries to be monitored and stress tested regularly over the next decade may rise to between 50 and 60. Stress testing should also be applied periodically to global systems that meet critical needs, including food supply systems, global public health systems, supply chains for critical materials, and disaster relief systems, as well as to international emergency response systems.

Decision science techniques should be used and further developed to ensure that the stress tests make the best use of the available information. Stress testing might draw on various methods, including qualitative interpretation of available knowledge, formal modeling, and interactive gaming approaches. Research analysts, area experts, and others might contribute in various ways, such as conducting analyses, developing models, and playing roles in gaming exercises. Decision science techniques should be employed to design the processes and interpret the input from different kinds of expertise and modes of analysis in order to make the best possible use of information. The stress-testing exercises should themselves be monitored and critically evaluated so that stress-testing methods can be improved over time.

References

2030 Water Resources Group. 2009. *Charting our water future: Economic frameworks to inform decision-making*. West Perth, Australia: 2030 Water Resources Group.
Abbott, P.C., C. Hurt, and W.E. Tyner. 2008. *What's driving food prices?* Farm Foundation. Available: http://www.farmfoundation.org/news/articlefiles/404-FINAL%20WDFP%20REPORT%207-28-08.pdf (accessed August 4, 2012).
Abbott, P.C., C. Hurt, and W.E. Tyner. 2011. *What's driving food prices in 2011?* Farm Foundation. Available: http://www.farmfoundation.org/news/articlefiles/1742-FoodPrices_web.pdf (accessed August 4, 2012).
Achen, C.H., and L.M. Bartels. 2004. *Blind retrospection: Electoral responses to drought, flu, and shark attacks*. Princeton, NJ: Princeton University.
Adams, G., L.T. O'Brien, and J.C. Nelson. 2006. Perceptions of racism in Hurricane Katrina: A liberation psychology analysis. *Analyses of Social Issues and Public Policy* 6(1):215–235.
Adano, W.R., T. Dietz, K. Witsenburg, and F. Zaal. 2012. Climate change, violent conflict, and local institutions in Kenya's drylands. *Journal of Peace Research* 49(1):65–80.
Adger, W.N. 2003. Social capital, collective action, and adaptation to climate change. *Economic Geography* 79(4):387–404.
Adger, W.N. 2006. Vulnerability. *Global Environmental Change* 16:268–281.
Adger, W.N., T.P. Hughes, C. Folke, S.R. Carpenter, and J. Rockström. 2005. Social–ecological resilience to coastal disasters. *Science* 309(5737):1,036–1,039.
Adger, W.N., H. Eakin, and A. Winkels. 2009a. Nested and teleconnected vulnerabilities to environmental change. *Frontiers in Ecology and the Environment* 7:150–157.
Adger, W.N., I. Lorenzoni, and K.L. O'Brien. 2009b. *Adapting to climate change: Thresholds, values, governance*. Cambridge, UK: Cambridge University Press.
Afifi, T. 2011. Economic or environmental migration? The push factors in Niger. *International Migration* 49:e95–e124.
Agarwal, B. 1995. Conceptualizing environmental collective action: Why gender matters. *Cambridge Journal of Economics* 24:283–310.
Alam, E., and A.E. Collins. 2010. Cyclone disaster vulnerability and response experiences in coastal Bangladesh. *Disasters* 34(4):931–954.

Aldrich, D.P. 2012. *Building resilience: Social capital in post-disaster recovery.* Chicago, IL: University of Chicago Press.

Alesina, A., S. Özler, N. Roubini, and P. Swagel. 1996. Political instability and economic growth. *Journal of Economic Growth* 1(2):189–211.

Alley, R.B., J. Martozke, W.D. Nordhaus, J.T. Overpeck, D.M. Peteet, R.A. Pielke Jr., R.T. Pierrehumbert, P.B. Rhines, T.F. Stocker, L.D. Talley, and J.M. Wallace. 2003. Abrupt climate change. *Science* 299:2005–2010.

Amery, H.A. 2002. Water wars in the Middle East: A looming threat. *Geographical Journal* 168(4):313–323.

Ansart, S., C. Pelat, P.-Y. Boelle, F. Carrat, A. Flahault, and A.-J. Valleron. 2009. Mortality burden of the 1918–1919 influenza pandemic in Europe. *Influenza and Other Respiratory Viruses* 3(3):99–106.

Archer, D.R., and H.J. Fowler. 2004. Spatial and temporal variations in precipitation in the Upper Indus Basin, global teleconnections and hydrological implications. *Hydrology and Earth System Sciences* 8(1):47–61.

Arora-Jonsson, S. 2011. Virtue and vulnerability: Discourses on women, gender, and climate change. *Global Environmental Change* 21:744–751.

Australian Government. 2009. *Metrological aspects of the 7 February 2009 Victorian fires, an overview.* Bureau of Meteorology Report for the 2009 Victorian Bushfires Royal Commission. Available: http://www.royalcommission.vic.gov.au/getdoc/f1eaba2f-414f-4b24-bcda-d81ff680a061/WIT.013.001.0012.pdf (accessed October 11, 2012).

Ayub, I. 2011. Police seek help over "daily" power riots. *Dawn*, May 20. Available: http://www.dawn.com/2011/05/20/police-seek-help-over-daily-power-riots.html (accessed October 4, 2012).

Bankoff, G., D. Hilhorst, and G. Frerks. 2004. *Mapping vulnerability: Disasters, development and people.* London: Earthscan.

Barnett, E.D. 2007. Yellow fever: Epidemiology and prevention. *Clinical Infectious Diseases* 44(6):850–856.

Barnett, J. 2006. Climate change, insecurity, and justice. Pp. 115–130 in *Fairness in adaptation to climate change.* W.N. Adger, J. Paavola, M.J. Mace, and S. Huq, Eds. Cambridge, MA: MIT Press.

Barnett, J., and W.N. Adger. 2007. Climate change, human security and violent conflict. *Political Geography* 26(6):639–655.

Barrett, A.D., and S. Higgs. 2007. Yellow fever: A disease that has yet to be conquered. *Annual Review of Entomology* 52:209–229.

Barry, J.M. 2005. *The great influenza: The story of the deadliest pandemic in history.* New York: Penguin Books.

Bates, R. 2008. State failure. *Annual Review of Political Science* 11:1–12.

Bellemare, M. 2012. *Rising food prices, food price volatility, and political unrest.* Duke University. Available: http://mpra.ub.uni-muenchen.de/31888/1/BellemareFoodPrices June2011.pdf (accessed August 5, 2012).

Benavides-Solorio, J. de D., and L.H. MacDonald. 2005. Measurement and prediction of post-fire erosion at the hillslope scale, Colorado Front Range. *International Journal of Wildland Fire* 14:1–18.

Benson, C., and E.J. Clay. 2004. *Understanding the economic and financial impacts of natural disasters.* Washington, DC: World Bank.

Bentz, B. 2008. *Western U.S. bark beetles and climate change.* Washington, DC: U.S. Department of Agriculture, Forest Service, Climate Change Resource Center. Available: http://www.fs.fed.us/ccrc/topics/bark-beetles.shtml (accessed August 5, 2012).

Bergholt, D., and P. Lujala. 2012. Climate related natural disasters, economic growth, and armed civil conflict. *Journal of Peace Research* 49(1):147–162.

Bernauer, T., and T. Siegfried. 2012. Climate change and international water conflict in Central Asia. *Journal of Peace Research* 49(1):227-240.
Bernauer, T., T. Böhmelt, and V. Koubi. 2012. Environmental changes and violent conflict. *Environmental Research Letters* 7(1):1-8.
Biermann, F., and I. Boas. 2010. Preparing for a warmer world: Towards a global governance system to protect climate refugees. *Global Environmental Politics* 10(1):60-88.
Bintanja, R., R.S.W. van de Wal, and J. Oerlemans. 2005. Modelled atmospheric temperatures and global sea levels over the past million years. *Nature* 437:125-128.
Black, R, W.N. Adger, N.W. Arnell, S. Dercon, A. Geddes, and D.S.G. Thomas. 2011a. Migration and global environmental change. *Global Environmental Change* 21S:S1-S2.
Black, R., W.N. Adger, N.W. Arnell, S. Dercon, A. Geddes, and D.S.G. Thomas. 2011b. The effect of environmental change on human migration. *Global Environmental Change* 21(Suppl. 1):S3-S11.
Black, R., S. Bennett, S. Thomas, and R. Beddington. 2011c. Climate change: Adaptation as migration. *Nature* 478:447-449.
Blaikie, P., T. Cannon, I. Davis, and B. Wisner. 1994. *At risk: Natural hazards, people's vulnerability, and disasters*. London, UK: Routledge.
Boëthius, G. 2012. Forging the ties the bind: Comparing the factors behind electricity market integration in the EU and ASEAN. EUC Working Paper No. 6. Singapore: EU Centre in Singapore.
Bolch, T., A. Kulkarni, A. Kääb, C. Huggel, F. Paul, J.G. Cogley, H. Frey, J.S. Kargel, K. Fujita, M. Scheel, S. Bajracharya, and M. Stoffel. 2012. The state and fate of Himalayan glaciers. *Science* 336(6079):310-314.
Brady, H.E., and D. Collier. (Eds.). 2010. *Rethinking social inquiry: Diverse tools, shared standards, 2nd ed*. Lanham, MD: Rowman and Littlefield.
Brancati, D. 2007. Political aftershocks: The impacts of earthquakes on intrastate conflict. *Journal of Conflict Resolution* 51(5):715-743.
Breisinger, C., O. Ecker, and P. Al-Riffai. 2011. *Economics of the Arab Awakening: From revolution to transformation and food security*. IFPRI Policy Brief 18. Washington, DC: International Food Policy Research Institute.
Brklacich, M., M. Chazan, and H.G. Bohle. 2010. Human security, vulnerability, and global environmental change. Pp. 35-52 in *Global environmental change and human security*. R. Matthew, J. Barnett, B. McDonald, and K. O'Brien, Eds. Cambridge, MA: MIT Press.
Brooks, N., W.N. Adger, and P.M. Kelly. 2005. The determinants of vulnerability and adaptive capacity at the national level and the implications for adaptation. *Global Environmental Change* 15:151-163.
Brown, L.R. 2011. When the Nile runs dry. *New York Times*. June 1.
Buhaug, H., N.P. Gleditsch, and O.M. Theisen. 2010. Implications of climate change for armed conflict. Pp. 75-102 in *Social dimensions of climate change: Equity and vulnerability in a warming world*. R. Mearns and A. Norton, Eds. Washington, DC: World Bank.
Bulloch, J., and A. Darwish. 1993. *Water wars: Coming conflicts in the Middle East*. London, UK: St. Edmundsbury Press.
Burkett, V. 2011. Global climate change implications for coastal and offshore oil and gas development. *Energy Policy* 39(12):7719-7725.
Busby, J.W. 2007. *Climate change and national security: An agenda for action*. New York: Council on Foreign Relations.
Butler, C.K., and S. Gates. 2012. African range wars: Climate, conflict, and property rights. *Journal of Peace Research* 49(1):23-34.
Butzer, K.W. 2012. Collapse, environment, and society. *Proceedings of the National Academy of Sciences* 109:3628-3631.

Butzer, K.W., and G.H. Endfield. 2012. Critical perspectives on historical collapse. *Proceedings of the National Academy of Sciences* 109:3628-3631.

Cannon, T. 1993. A hazard need not a disaster make: Vulnerability and the causes of "natural disasters." Pp. 92-105 in *Natural disasters: Protecting vulnerable communities*. P.A. Merriman and C.W.A. Browitt, Eds. New York: Thomas Telford.

Cannon, T. 1994. Vulnerability analysis and the explanation of "natural" disasters. Pp. 13-30 in *Disasters, development, and environment*. A. Varley, Ed. New York: John Wiley and Sons.

Cannon, T. 2000. Vulnerability analysis and disasters. Pp. 45-55 in *Floods*. D.J. Parker, Ed. New York: Routledge.

Carmen, H.E., C. Parthemore, and W. Rogers. 2010. *Broadening horizons: Climate change and the U.S. Armed Forces*. Washington, DC: Center for a New American Security.

Carroll, A.L., S.W. Taylor, J. Regniere, and L. Safranyik. 2003. Effect of climate change on range expansion by the mountain pine beetle in British Columbia. Pp. 223-232 in *Mountain pine beetle symposium: Challenges and solutions*. Information Report No. BC-X-399. T.L. Shore, J.E. Brooks, and J.E. Stone, Eds. Victoria, Canada: Natural Resources Canada, Canadian Forest Service, Pacific Forestry Centre.

Carter, M., P. Little, T. Mogues, and W. Negatu. 2007. Poverty traps and natural disasters in Ethiopia and Honduras. *World Development* 35(5):835-856.

Center for Climate and Energy Solutions. 2009. *National security implications of global climate change*. Arlington, VA: Center for Climate and Energy Solutions.

Center for Naval Analysis. 2007. *National security and the threat of climate change*. Alexandria, VA: CNA Corporation.

Centers for Disease Control and Prevention. 2009. *Seasonal influenza*. Available: http://www.cdc.gov/flu/ (accessed October 4, 2012).

Chatterji, M., S. Arlosoroff, and G. Guha. (Eds.). 2002. *Conflict management of water resources*. Burlington, VT: Ashgate.

Chen, P.-S., F.T. Tsai, C.K. Lin, C.-Y. Yang, C.-C. Chan, C.-Y. Young, and C.-H. Lee. 2010. Ambient influenza and avian influenza virus during dust storm days and background days. *Environmental Health Perspectives* 118(9):1211-1216.

Ciezadlo, A. 2011. Let them eat bread: how food subsidies prevent (and provoke) revolutions in the Middle East. Pp. 229-235 in *The new Arab revolt: What happened, what it means, and what comes next*. New York: Council on Foreign Relations.

Clinton, W.J. 1996. *Presidential decision directive NSTC-7*. Washington, DC: The White House.

Collins, A.E. 2011. *Disaster and development*. New York: Routledge.

Collins, M. 2000. Understanding uncertainties in the response of ENSO to greenhouse warming. *Geophysical Research Letters* 27(21):3,500-3,512.

Confalonieri, U., B. Menne, R. Akhtar, K.L. Ebi, M. Hauengue, R.S. Kovats, B. Revich, and A. Woodward. 2007. Human health. Pp. 391-431 in *Climate change 2007: Impacts, adaptation and vulnerability*. Contribution of Working Group II to the Fourth Assessment Report of the Intergovernmental Panel on Climate Change. M.L. Parry, O.F. Canziani, J.P. Palutikof, P.J. van der Linden, and C.E. Hanson, Eds. Cambridge, UK: Cambridge University Press.

Connor, D. 2012. Hard drive prices rise due to Thai floods. *InformationWeek*. Available: http://www.informationweek.com/storage/data-protection/hard-drive-prices-rise-due-to-thai-flood/232301534 (accessed October 4, 2012).

Cooley, J.K. 1984. The war over water. *Foreign Policy* 54:3-26.

Coppola, D.P. 2011. *Introduction to international disaster management, 2nd ed*. Oxford, UK: Butterworth-Heinemann.

Cuny, F.C. 1983. *Disasters and development*. New York: Oxford University Press.

Cutter, S., and C. Finch. 2008. Temporal and spatial changes in social vulnerability to natural hazards. *Proceedings of the National Academy of Sciences* 105:2,301–2,306.
Cutter, S.L., B.J. Boruff, and W.L. Shirley. 2003. Social vulnerability to environmental hazards. *Social Science Quarterly* 84(2):242–261.
Dai, A., K.E. Trenberth, and T.R. Karl. 1998. Global variations in droughts and wet spells: 1900–1995. *Geophysical Research Letters* 25(17):3,367–3,370.
Davis, M. 2002. *Late Victorian holocausts: El Niño famines and the making of the third world*. New York: Verso.
Dawe, D. 2008. Have recent increases in international cereal prices been transmitted to domestic economies? The experience in seven large Asian countries. ESA Working Paper No. 08-03. Food and Agriculture Organization. Available: ftp://ftp.fao.org/es/esa/esawp/ESAWP_08_03.pdf (accessed August 10, 2012).
Dawn. 2011. IRSA urges provinces to conserve water. *Dawn*, July 18. Available: http://dawn.com/2011/07/18/irsa-urges-provinces-to-conserve-water/ (accessed August 5, 2012).
De Graaf, R., and R. Van Der Brugge. 2010. Transforming water infrastructure by linking water management and urban renewal in Rotterdam. *Technological Forecasting and Social Change* 77(8):1282–1291.
De Sherbinin, A., A. Schiller, and A. Pulsipher. 2007. The vulnerability of global cities to climate hazards. *Environment and Urbanization* 19(1):39–64.
De Stefano, L., J. Duncan, S. Dinar, K. Stahl, K.M. Strzepek, and A.T. Wolf. 2012. Climate change and the institutional resilience of international river basins. *Journal of Peace Research* 49(1):193–209.
Defense Science Board. 2011. *Trends and implications of climate change for national and international security*. Washington, DC: Department of Defense.
Dell, M., B.F. Jones, and B.A. Olken. 2012. Temperature shocks and economic growth: Evidence from the last half century. *American Economic Journal: Macroeconomics*, 4(3):66–95.
Denton, F. 2002. Climate change vulnerability, impacts, and adaptation: Why does gender matter? *Gender and Development* 10(2):10–20.
Diamond, J.M. 2005. *Collapse: How societies choose to fail or succeed*. New York: Viking Press.
Drury, A.C., and R.S. Olson. 1998. Disasters and political unrest: An empirical investigation. *Journal of Contingencies and Crisis Management* 6:153–161.
du Prel, J.B., W. Puppe, B. Grondahl, M. Knuf, F. Weigl, F. Schaaf, and H.J. Schmitt. 2009. Are meteorological parameters associated with acute respiratory tract infections? *Clinical Infectious Diseases* 49(6):861–868.
Eakin, H., and A.L. Luers. 2006. Assessing the vulnerability of social–environmental systems. *Annual Review of Environment and Resources* 31:365–394.
Easterling, D.R., and M.F. Wehner. 2009. Is the climate warming or cooling? *Geophysical Research Letters* 36:L08706.
Elhance, A.P. 1999. *Hydropolitics in the third world: Conflict and cooperation in international river basins*. Washington, DC: U.S. Institute of Peace.
Enarson, E. 1998. Through women's eyes: A gendered research agenda for disaster social science. *Disasters* 22(2):157–173.
Enarson, E. 2012. *Women in disaster*. Nashville, TN: Vanderbilt University Press.
Express Tribune. 2012. Loadshedding: Protesters attack PML-Q MNA's house in Kamalia. *The Express Tribune*, June 19. Available: http://tribune.com.pk/story/395929/loadshedding-protesters-attack-mnas-house-in-kamalia/ (accessed August 6, 2012).
Fagan, B. 2009. *Floods, famines, and emperors: El Niño and the fate of civilizations*. New York: Basic Books.

Famine Early Warning Systems Network. 2008. *Famine Early Warning Systems Network (FEWS NET) markets and trade strategy for 2005–2010.* Available: http://www.fews.net/docs/special/FEWSNETMarketandTradeStrategy2005-2010.pdf (accessed July 10, 2012).
Feely, R.A., T. Takahashi, R. Wanninkhof, M.J. McPhaden, C.E. Cosca, S.C. Sutherland, and M.-E. Carr. 2006. Decadal variability of the air–sea CO_2 fluxes in the equatorial Pacific Ocean. *Journal of Geophysical Research* 111. doi:10.1029/2005JC003129.
Feitelson, E., A. Tamimi, and G. Rosenthal. 2012. Climate change and security in the Israeli–Palestinian conflict. *Journal of Peace Research* 49(1):241–257.
Finch, C., C.T. Emrich, and S.L. Cutter. 2010. Disaster disparities and differential recovery in New Orleans. *Population and Environment* 31(4):179–202.
Fingar, T. 2008. *National intelligence assessment on the national security implications of global climate change to 2030: Statement for the record of Dr. Thomas Fingar.* Before the Permanent Select Committee on Intelligence and the Select Committee on Energy Dependence and Global Warming, House of Representatives. Washington, DC. June 25. Available: http://media.npr.org/documents/2008/jun/warming_intelligence.pdf (accessed August 10, 2012).
Food and Agriculture Organization of the United Nations. 2012. *FAOSTAT.* Available: http://faostat.fao.org (accessed August 10, 2012).
Foresight. 2011. *Migration and global environmental change: Future challenges and opportunities.* Final project report. London, UK: Government Office for Science. Available: http://www.bis.gov.uk/assets/foresight/docs/migration/11-1116-migration-and-global-environmental-change.pdf (accessed October 4, 2012).
Foster, G., and S. Rahmstorf. 2011. Global temperature evolution 1979–2010. *Environmental Research Letters* 6(4). doi:10.1088/1748-9326/6/4/044022.
Fuller, T. 2011. Thailand flooding cripples hard-drive suppliers. *New York Times*, November 7.
Fuller, T. 2012. As Myanmar changes, so does its leader. *New York Times*, April 3.
Füssel, H.-M. 2007. Vulnerability: A generally applicable conceptual framework for climate change research. *Global Environmental Change* 17(2):155–167.
Gaillard, J. 2010. Vulnerability, capacity and resilience: Perspectives for climate and developmental policy. *Journal of International Development* 22:218–232.
Gardner, C.L., and K.D. Ryman. 2010. Yellow fever: A reemerging threat. *Clinics in Laboratory Medicine* 30(1):237–260.
Garrard-Burnett, V. 2009. Under God's thumb: The 1976 Guatemala earthquake. Chapter 6 in *Aftershocks: Earthquakes and popular politics in Latin America.* J. Buchenau and L.L. Johnson, Eds. Albuquerque, NM: University of New Mexico Press.
Gasper, J.T., and A. Reeves. 2011. Make it rain? Retrospection and the attentive electorate in the context of natural disasters. *American Journal of Political Science* 55(2):340–355.
Gawronski, V., and R.S. Olson. 2000. "Special" versus "normal" time corruption: An exploration of Mexican attitudes. *Cambridge Review of International Affairs* 14:345–361.
Gawronski, V., and R.S. Olson. 2013. Disasters as crisis triggers for critical junctures? The 1976 Guatemala case. *Latin American Politics and Society* 55(2).
Gemenne, F. 2011. Why the numbers don't add up: A review of estimates and predictions of people displaced by environmental changes. *Global Environmental Change* 21(Suppl. 1):S41–S49.
George, A.L. 1991. *Avoiding war: Problems of crisis management.* Boulder, CO: Westview Press.
George, A.L., and A. Bennett. 2005. *Case studies and theory development in the social sciences.* Cambridge, MA: MIT Press.

Ghumman, M. 2012a. IRSA prefers irrigation to power generation: 21 percent water shortage in Kharif estimated. *Business Recorder,* March 27. Available: http://www.brecorder.com/index.php?option=com_news&view=single&id=1169177 (accessed August 5, 2012).

Ghumman, M. 2012b. Power shortfall crosses 8,000mw mark. *Business Recorder,* June 17. Available: http://www.brecorder.com/top-news/1-front-top-news/62631-power-shortfall-crosses-8000mw-mark-.html (accessed August 5, 2012).

Giordano, M., M. Giordano, and A. Wolf. 2002. The geography of water conflict and cooperation: Internal pressures and international manifestations. *Geographical Journal* 168:293–312.

Gleason, K.L., J.H. Lawrimore, D.H. Levinson, T.R. Karl, and D.J. Karoly. 2008. A revised U.S. climate extremes index. *Journal of Climate* 21(10):2,124–2,137. Available: http://journals.ametsoc.org/doi/full/10.1175/2007JCLI1883.1 (accessed November 15, 2012).

Gleditsch, N.P. 2012. Whither the weather? Climate change and conflict. *Journal of Peace Research* 49(1):3–9.

Gleditsch, N.P., H. Buhaug, and O.M. Theisen. 2011. *Climate change and armed conflict*. Revised version of a paper prepared for the Department of Energy/Environmental Protection Agency Workshop on Research on Climate Change Impacts and Associated Economic Damages, January 27–28, Washington, DC.

Global Climate Observing System. 2010. *Implementation plan for the Global Climate Observing System for climate in support of the UNFCCC*. Geneva, Switzerland: World Meteorological Organization. Available: http://www.wmo.int/pages/prog/gcos/Publications/gcos-138.pdf (accessed August 6, 2012).

Goldstein, J.A. 2011. *Winning the war on war: The decline of armed conflict worldwide*. New York: Dutton.

Goldstone, J.A. 2012. *Forecasting political instability: PITF history and approach*. Briefing to the Committee on Assessing the Impacts of Climate Change on Social and Political Stresses, March 2, Washington, DC.

Goldstone, J.A., R.H. Bates, D.L. Epstein, R.T. Gurr, M.B. Lustik, M.G. Marshall, J. Ulfelder, and M. Woodward. 2010. A global model for forecasting political instability. *American Journal of Political Science* 54(1):190–208.

Gotham, K.F., and R. Campanella. 2011. Coupled vulnerability and resilience: The dynamics of cross-scale interactions in post-Katrina New Orleans. *Ecology and Society* 16(3):12.

Gould, E.A., and S. Higgs. 2009. Impact of climate change and other factors on emerging arbovirus diseases. *Transactions of the Royal Society of Tropical Medicine and Hygiene* 103(2):109–121.

Government of Pakistan. 2012. *Pakistan economic survey, 2011–2012*. Islamabad: Government of Pakistan, Ministry of Finance. Available: http://finance.gov.pk/survey_1112.html (accessed October 4, 2012).

Gupta, E. 2008. Oil vulnerability index of oil-importing countries. *Energy Policy* 36:1,195–1,211.

Hamilton, J.D. 2003. What is an oil shock? *Journal of Econometrics* 113(2):363–398.

Hamilton, J.D. 2008. Oil and the macroeconomy. In *The new Palgrave dictionary of economics, 2nd ed.* S. Durlauf and L. Blume, Eds. Basingstoke, UK: Palgrave Macmillan. Available: http://www.dictionaryofeconomics.com/article?id=pde2008_E000233 (accessed October 10, 2012)

Hansen, J.E., and M. Sato. 2012. Paleoclimate implications for human-made climate change. Pp. 21–48 in *Climate change: Inferences from paleoclimate and regional aspects*. A. Berger, F. Mesinger, and D. Sijack, Eds. Vienna, Austria: Springer.

Hansen, J., R. Ruedy, M. Sato, and K. Lo. 2010. Global surface temperature change. *Reviews of Geophysics* 48:RG4004. doi:10.1029/2010RG000345.

Hansen, J., M. Sato, and R. Ruedy. 2012. Perception of climate change. *Proceedings of the National Academy of Sciences* 109(37):E2415–E2423. Available: http://www.pnas.org/cgi/doi/10.1073/pnas.1205276109 (accessed October 4, 2012).
Hanson, S., R. Nicholls, N. Ranger, S. Hallegatte, J. Corfee-Morlot, C. Herweijer, and J. Chateau. 2011. A global ranking of port cities with high exposure to climate extremes. *Climatic Change* 104:89–111.
Haque, U., M. Hashizume, K.N. Kolivras, H.J. Overgaard, B. Das, and T. Yamamoto. 2012. Reduced death rate from cyclones in Bangladesh: What more needs to be done? *Bulletin of the World Health Organization* 90(2):150–156.
Hazards and Vulnerability Research Institute. 2012. *Social vulnerability index for the United States: 2006–2012.* Department of Geography, University of South Carolina. Available: http://webra.cas.sc.edu/hvri/products/sovi.aspx (accessed October 15, 2012).
Healy, A., and N. Malhotra. 2009. Myopic voters and natural disaster policy. *American Political Science Review* 103(3):387–406.
Hendrix, C.S., and I. Salehyan. 2012. Climate change, rainfall, and social conflict in Africa. *Journal of Peace Research* 49(1):35–50.
Herweijer, C., and R. Seager. 2008. The global footprint of persistent extra-tropical drought in the instrumental era. *International Journal of Climatology* 28(13):1,761–1,774. DOI:10.1002/joc.1590. Available: http://www.ldeo.columbia.edu/res/div/ocp/pub/herweijer/Herweijer_Seager_IJC.pdf (accessed October 4, 2012).
Hewitt, J.J., J. Wilkenfeld, and T.R. Gurr. 2012. *Peace and conflict 2012.* Boulder, CO: Paradigm.
Hewitt, K. 2005. The Karakoram anomaly? Glacier expansion and the "elevation effect," Karakoram Himalaya. *Mountain Research and Development* 25(4):332–340.
Hodell, D.A., J.H. Curtis, and M. Brenner. 1995. Possible role of climate in the collapse of Classic Maya civilization. *Nature* 375:391–394.
Homer-Dixon, T. 1991. On the threshold: Environmental changes as causes of acute conflict. *International Security* 16(2):76–116.
Homer-Dixon, T. 1994. Environmental scarcities and violent conflict: Evidence from cases. *International Security* 19(1):5–40.
Homer-Dixon, T. 1996. Strategies for studying causation in complex ecological–political systems. *Journal of Environmental Development* 5(2):132–148.
Homer-Dixon, T. 1999. *Environment, scarcity, and violence.* Princeton, NJ: Princeton University Press.
Homer-Dixon, T. 2007. Terror in the weather forecast. *New York Times,* April 24.
Hsiang, S.M., K.T. Meng, and M.A. Cane. 2011. Civil conflicts are associated with the global climate. *Nature* 476:438–441.
Hugo, G. 2011. Future demographic change and its interactions with migration and climate change. *Global Environmental Change* 21(Suppl. 1):S21–S33.
Hurrell, J.W., G.A. Meehl, D. Bader, T.L. Delworth, B. Kirtman, and B. Wielicki. 2009. A unified modeling approach to climate system prediction. *Bulletin of the American Meteorological Society* 90(12):1819–1832.
Hurrell, J.W., T.L. Delworth, G. Danagasoglu, H. Drange, K. Drinkwater, S. Griffies, N.J. Holbrook, B. Kirtman, N. Keenlyside, M. Latif, J. Marotzke, J. Murphy, G.A. Meehl, T. Palmer, H. Pohlmann, T. Rosati, R. Seager, D. Smith, R. Sutton, A. Timmermann, K.E. Trenberth, J. Tribbia, and M. Visbeck. 2010. Decadal climate prediction: Opportunities and challenges. *Proceedings of OceanObs'09.* Available: https://abstracts.congrex.com/scripts/jmevent/abstracts/FCXNL-09A02a-1661836-1-cwp3b03.pdf (accessed October 4, 2012).

Ianchovichina, E., J. Ludger-Loening, and C.A. Wood. 2012. *How vulnerable are Arab countries to global food price shocks?* World Bank Working Paper No. 6018. Available: http://papers.ssrn.com/sol3/papers.cfm?abstract_id=2031389 (accessed October 4, 2012).
Intergovernmental Panel on Climate Change. 2007. *Climate change 2007: Impacts, adaptation, and vulnerability.* Cambridge, UK: Cambridge University Press.
Intergovernmental Panel on Climate Change. 2012. *Managing the risks of extreme events and disasters to advance climate change adaptation.* Special report of working groups I and II of the Intergovernmental Panel on Climate Change. C.B. Field, V. Barros, T.F. Stocker, D. Qin, D.J. Dokken, K.L. Ebi, M.D. Mastrandrea, K.J. Mach, G.-K. Plattner, S.K. Allen, M. Tignor, and P.M. Midgley, Eds. Cambridge, UK: Cambridge University Press.
International Institute for Strategic Studies. 2011. *The IISS transatlantic dialogue on climate change and security: Report to the European Commission.* London, UK: International Institute for Strategic Studies.
Jafrani, N. 2010. *Pakistani water crisis: Concerns about regional stability and conflict.* Center for Strategic and International Studies. May 10. Available: http://csis.org/blog/pakistani-water-crisis-concerns-about-regional-stability-and-conflict (accessed August 10, 2012).
Jamasb, T., and M.G. Pollitt. 2005. *Electricity market reform in the European Union: Review of progress towards liberalisation and integration.* Working Paper No. 05-003. Cambridge, MA: MIT Center for Energy and Environmental Policy Research.
Johnson, S. 2011. *Climate and catastrophe in Cuba and the Atlantic world in the age of revolution.* Chapel Hill, NC: University of North Carolina Press.
Johnstone, S., and J. Mazo. 2011. Global warming and the Arab Spring. *Survival* 53(2):11–17.
Jones, C.D., M. Collins, P.M. Cox, and S.A. Spall. 2001. The carbon cycle response to ENSO: A coupled climate–carbon cycle model study. *Journal of Climate* 14(21):4113–4129.
Juneja, S. 2008. *Disasters and poverty: The risk nexus: A review of literature.* Geneva, Switzerland: United Nations International Strategy for Disaster Reduction. Available: http://ebookbrowse.com/juneja-disasters-and-poverty-the-risk-nexus-doc-d191326079 (accessed August 10, 2012).
Kahl, C.H. 2006. *States, scarcity, and civil strife in the developing world.* Princeton, NJ: Princeton University Press.
Kahn, M.E. 2005. The death toll from natural disasters: The role of income, geography, and institutions. *Review of Economics and Statistics* 87(2):271–284.
Kasperson, R.E., and J.X. Kasperson. 2001. *Climate change, vulnerability and social justice.* Stockholm, Sweden: Stockholm Environment Institute.
Keesing, F., L.K. Belden, P. Daszak, A. Dobson, C.D. Harvell, R.D. Holt, P. Hudson, A. Jolles, K.E. Jones, C.E. Mitchell, S.S. Myers, T. Bogich, and R.S. Ostfeld. 2010. Impacts of biodiversity on the emergence and transmission of infectious diseases. *Nature* 468:647–652.
Keller, E.J. 1992. Drought, war, and the politics of famine in Ethiopia and Eritrea. *Journal of Modern African Studies* 30(4):609–624.
Keskitalo, E. 2009. Governance in vulnerability assessment: The role of globalising decision-making networks in determining local vulnerability and adaptive capacity. *Mitigation and Adaptation Strategies for Global Change* 14(2):185–201.
Kim, N. 2012. How much more exposed are the poor to natural disasters? Global and regional measurement. *Disasters* 36(2):195–211.
King, G., R.O. Keohane, and S. Verba. 1994. *Designing social inquiry: Scientific inference in qualitative research.* Princeton, NJ: Princeton University Press.
Koubi, V., T. Bernauer, A. Kalbhenn, and G. Spilker. 2012. Climate variability, economic growth, and civil conflict. *Journal of Peace Research* 49(1):113–127.
Kovats, R.S., and S. Hajat. 2008. Heat stress and public health: A critical review. *Annual Review of Public Health* 29:41–55.

Lagi, M., K.Z. Bertrand, and Y. Bar-Yam. 2011. The food crises and political instability in North Africa and the Middle East. Available: http://arxiv.org/pdf/1108.2455v1.pdf (accessed October 4, 2012).

Latif, M., T.L. Delworth, D. Dommenget, H. Drange, W. Hazeleger, J.W. Hurrell, N. Keenlyside, G.A. Meehl, and R.T. Sutton. 2010. Dynamics of decadal variability and implications for its prediction. *Proceedings of OceanObs'09: Sustained Ocean Observations and Information for Society*. Available: http://www.oceanobs09.net/proceedings/cwp/Latif-OceanObs09.cwp.53.pdf (accessed October 4, 2012).

Lau, W.K.M., and K.-M. Kim. 2012. The 2010 Pakistan flood and Russian heat wave: Teleconnection of hydrometeorological extremes. *Journal of Hydrometeorology* 13:392–403.

Lee, J.R. 2009. *Climate change and armed conflict: Hot and cold wars*. New York: Routledge.

Leichenko, R.M., and K.L. O'Brien. 2008. *Environmental change and globalization: Double exposures*. New York: Oxford University Press.

Lennon, A.T.J., J. Gulledge, J.R. McNeill, J. Podesta, P. Ogden, L. Fuerth, R.J. Woolsey, J. Smith, R. Weitz, and D. Mix. 2007. *The age of consequences: The foreign policy and national security implications of global climate change*. Washington, DC: Center for Strategic and International Studies.

Lenton, T.M., H. Held, E. Kriegler, J.W. Hall, W. Lucht, S. Rahmstorf, and H.J. Schellnhuber. 2008. Tipping elements in the Earth's climate system. *Proceedings of the National Academy of Sciences* 105:1,786–1,793.

Levy, J.S., and W.R. Thompson. 2010. *Causes of war*. Chichester, UK: John Wiley and Sons.

Lilleør, H.B., and K. Van den Broeck. 2011. Economic drivers of migration and climate change in LDCs. *Global Environmental Change* 21(Suppl. 1):S70–S81.

Lindgren, E., Y. Andersson, J.E. Suk, B. Sudre, and J.C. Semenza. 2012. Monitoring EU emerging infectious disease risk due to climate change. *Science* 336:418–419.

Liverman, D.M. 1990. Drought impacts in Mexico: Climate, agriculture, technology, and land tenure in Sonora and Puebla. *Annals of the Association of American Geographers* 80(1):49–72.

London Climate Change Partnership. 2006. *Adapting to climate change: Lessons from London*. London, UK: Greater London Authority.

Malhotra, N., and A.G. Kuo. 2008. Attributing blame: The public's response to Hurricane Katrina. *Journal of Politics* 70(1):120–135.

Marten, K. 2010. Failing states and conflict. Pp. 2012–2022 in *The international studies encyclopedia*. R.A. Denemark, Ed. Malden, MA: Wiley-Blackwell.

Marty, F. 2001. *Managing international rivers: Problems, politics, and institutions*. Bern, Switzerland: Peter Lang.

Mastrandrea, M.D., C.B. Field, T.F. Stocker, O. Edenhofer, K.L. Ebi, D.J. Frame, H. Held, E. Kriegler, K.J. Mach, P.R. Matschoss, G.-K. Plattner, G.W. Yohe, and F.W. Zwiers. 2010. *Guidance note for lead authors of the IPCC Fifth Assessment Report on Consistent Treatment of Uncertainties*. Intergovernmental Panel on Climate Change. Available: http://www.ipcc.ch/pdf/supporting-material/uncertainty-guidance-note.pdf (accessed September 24, 2012).

McCalla, R.B. 1992. *Uncertain perceptions: U.S. Cold War crisis decision making*. Ann Arbor: University of Michigan Press.

McLeman, R., and B. Smit. 2006. Migration as an adaptation to climate change. *Climatic Change* 76:31–53.

Mearian, L. 2011. Impact of hard drive shortage to linger through 2013. *Computerworld*. December 9. Available: http://www.computerworld.com/s/article/9222522/Impact_of_hard_drive_shortage_to_linger_through_2013 (accessed October 4, 2012).

REFERENCES

Meehl, G.A., T.F. Stocker, W.D. Collins, P. Friedlingstien, A.T. Gaye, J.M. Gregory, A. Kitoh, R. Knutti, J.M. Murphy, A. Noda, S.C.B. Raper, I.G. Watterson, A.J. Weaver, and Z.-C. Zhao. 2007. Global climate projections. Pp. 747–845 in *Climate Change 2007. The physical science basis*. Contribution of Working Group I to the Fourth Assessment Report of the Intergovernmental Panel on Climate Change. S. Solomon, D. Qin, M. Manning, S.Z. Chen, M. Marquis, K.B. Averyt, M. Tignor, and H.L. Miller, Eds. Cambridge, UK: Cambridge University Press.

Meehl, G.A., C. Tebaldi, G. Walton, D. Easterling, and L. McDaniel. 2009. Relative increase of record high maximum temperatures compared to record low minimum temperatures in the U.S. *Geophysical Research Letters* 36:L23701–L23705.

Meehl, G.A., J.M. Arblaster, J.T. Fasullo, A. Hu, and K.E. Trenberth. 2011. Model-based evidence of deep-ocean heat uptake during surface-temperature hiatus periods. *Nature Climate Change* 1(7):360–364.

Michel, D. 2009. A river runs through it: Climate change, security challenges, and shared water resources. Pp. 73–103 in *Troubled waters: Climate change, hydropolitics, and transboundary resources*. D. Michel and A. Pandya, Eds. Washington, DC: Henry L. Stimson Center.

Munich Reinsurance Company. 2012. *Natural catastrophes 2011*. Available: http://www.munichre.com/app_pages/www/@res/pdf/media_relations/press_releases/2012/2012_01_04_munich_re_natural-catastrophes-2011_en.pdf?2 (accessed September 27, 2012).

Murray, V., G. McBean, M. Bhatt, S. Borsch, T.S. Cheong, W.F. Erian, S. Llosa, F. Nadim, M. Nunez, R. Oyun, and A.G. Suarez. 2012: Case studies. Pp. 487–542 in *Managing the risks of extreme events and disasters to advance climate change adaptation*. A special report of working groups I and II of the Intergovernmental Panel on Climate Change. C.B. Field, V. Barros, T.F. Stocker, D. Qin, D.J. Dokken, K.L. Ebi, M.D. Mastrandrea, K.J. Mach, G.-K. Plattner, S.K. Allen, M. Tignor, and P.M. Midgley, Eds. Cambridge, UK: Cambridge University Press.

NASA Goddard Institute for Space Studies. 2012. *Datasets and images. Global annual mean surface air temperature change*. Available: http://data.giss.nasa.gov/gistemp/graphs_v3/ (accessed August 15, 2012).

National Aeronautics and Space Administration. 2012. *GISS surface temperature analysis*. Available: http://data.giss.nasa.gov/gistemp/tabledata_v3/GLB.Ts+dSST.txt (accessed September 20, 2012).

National Research Council. 1992. *Global environmental change: Understanding the human dimensions*. Committee on the Human Dimensions of Global Change. P.C. Stern, O.R. Young, and D. Druckman, Eds. Washington, DC: National Academy Press.

National Research Council. 1999. *Human dimensions of global environmental change: Research pathways for the next decade*. Committee on the Human Dimensions of Global Change and Committee on Global Change Research. Washington, DC: National Academy Press.

National Research Council. 2001. *Under the weather: Climate, ecosystems, and infectious diseases*. Committee on Climate, Ecosystems, Infectious Diseases, and Human Health. Washington, DC: National Academy Press.

National Research Council. 2002. *Abrupt climate change: Inevitable surprises*. Washington, DC: National Academy Press.

National Research Council. 2006. *Facing hazards and disasters: Understanding human dimensions*. Committee on Disaster Research in the Social Sciences: Future Challenges and Opportunities. Washington, DC: The National Academies Press.

National Research Council. 2009. *Informing decisions in a changing climate*. Panel on Strategies and Methods for Climate-Related Decision Support. Washington, DC: The National Academies Press.

National Research Council. 2010a. *Advancing the science of climate change.* Panel on Advancing the Science of Climate Change. Washington, DC: The National Academies Press.
National Research Council. 2010b. *Monitoring climate change impacts: Metrics at the intersection of the human and earth systems.* Committee on Indicators for Understanding Global Climate Change. Washington, DC: The National Academies Press.
National Research Council. 2010c. *Adapting to the impacts of climate change.* Panel on Adapting to the Impacts of Climate Change. Washington, DC: The National Academies Press.
National Research Council. 2010d. *Verifying greenhouse gas emissions: Methods to support international climate agreements.* Washington, DC: The National Academies Press.
National Research Council. 2011a. *Climate stabilization targets: Emissions, concentrations, and impacts for decades to millennia.* Washington, DC: The National Academies Press.
National Research Council. 2011b. *National security implications of climate change for U.S. naval forces.* Washington, DC: The National Academies Press.
National Research Council. 2012a. *A review of the U.S. Global Change Research Program's strategic plan.* Committee to Advise the U.S. Global Change Research Program. Washington, DC: The National Academies Press.
National Research Council. 2012b. *Climate change: Evidence, impacts, and choices.* Washington, DC: The National Academies Press.
National Research Council. 2012c. *Himalayan glaciers: Climate change, water resources, and water security.* Washington, DC: The National Academies Press.
Natsios, A.S. 1995. NGOs and the UN system in complex humanitarian emergencies: Conflict or cooperation? *Third World Quarterly* 16(3):405–419.
Naylor, R.L., and W.P. Falcon. 2010. Food security in an era of economic volatility. *Population and Development Review* 36(4):693–723.
Nel, P., and M. Righarts. 2008. Natural disaster and the risk of violent civil conflict. *International Studies Quarterly* 52(1):159–185.
Neumeyer, E., and F. Barthel. 2011. Normalizing economic loss from natural disasters: A global analysis. *Global Environmental Change* 21:13–24.
Neumeyer, E., and T. Plümper. 2007. The gendered nature of natural disasters: The impact of catastrophic events on the gender gap in life expectancy, 1981–2002. *Annals of the Association of American Geographers* 97(3):551–566.
News International. 2012. Faisalabad traders launch civil disobedience movement. *News International,* April 25.
Nizamani, A., F. Rauf, and A.H. Khoso. 1998. Case study: Pakistan: population and water resources. In *Water and population dynamics: Case studies and policy implication.* A. de Sherbinin and V. Dompka, Eds. New York: American Association for the Advancement of Science. Available: http://www.aaas.org/international/ehn/waterpop/paki.htm (accessed October 4, 2012).
Nordås, R., and N.P. Gleditsch. 2007. Climate change and conflict. *Political Geography* 26(6):627–638.
Nur, A., and D. Burgess. 2008. *Apocalypse: Earthquakes, archaeology, and the wrath of God.* Princeton, NJ: Princeton University Press.
Ó Gráda, C. 2009. *Famine: A short history.* Princeton, NJ: Princeton University Press.
Ó Gráda, C. 2011. Famines past, famine's future. *Development and Change* 42(1):49–69.
O'Brien, K., and R. Leichenko. 2007. *Human security, vulnerability, and sustainable adaptation.* New York: United Nations Development Programme. Available: http://hdr.undp.org/en/reports/global/hdr2007-8/papers/O'Brien_Karen%20and%20Leichenko_Robin.pdf (accessed August 10, 2012).

O'Brien, K., R. Leichenko, U. Kelkar, and H. Venema. 2004. Mapping vulnerability to multiple stressors: Climate change and globalization in India. *Global Environmental Change* 14(4):303-313.
O'Brien, K.L., B. Hayward, and F. Berkes. 2009. Rethinking social contracts: Building resilience in a changing climate. *Ecology and Society* 14(2):12.
Office of the Director of National Intelligence. 2012. *Global water security: Intelligence community assessment*. Washington, DC: Office of the Director of National Intelligence.
Olson, R.S. 2000. Toward a politics of disaster: Losses, values, agendas, and blame. *International Journal of Mass Emergencies and Disasters* 18:265-287.
Olson, R.S., and V.T. Gawronski. 2003. Disasters as critical junctures? Managua, Nicaragua 1972 and Mexico City 1985. *International Journal of Mass Emergencies and Disasters* 21(1):5-35.
Olson, R.S., and V.T. Gawronski. 2010. From disaster event to political crisis: A "5C+A" framework for analysis. *International Studies Perspectives* 11(3):1-17.
Omelicheva, M. 2011. Natural disasters: Triggers of political instability? *International Interactions* 37(4):441-465.
Osman-Elasha, B. 2009. Women . . . In the shadow of climate change. *UN Chronicle* 46(3/4):54-55.
Paavola, J. 2008. Livelihoods, vulnerability and adaptation to climate change in Morogoro, Tanzania. *Environmental Science and Policy* 11:642-654.
Pakistan Institute of Legislative Development and Transparency. 2011. *Inter-provincial water issues in Pakistan*. Draft background paper. January. Available: http://www.pildat.org/publications/publication/WaterR/Inter-ProvincialWaterIssuesinPakistan-BackgroundPaper.pdf (accessed September 25, 2012).
Pakistan Water and Power Development Authority. 2011. *Hydro potential in Pakistan*. Available: http://www.wapda.gov.pk/pdf/BroHydpwrPotialApril2011.pdf (accessed October 4, 2012).
Parliament of Victoria. 2010. *2009 Victorian Bushfires Royal Commission, summary report, Volume 2*. Government Printer for the State of Victoria, Melbourne, Australia. Available: http://www.royalcommission.vic.gov.au/finaldocuments/summary/PF/VBRC_Summary_PF.pdf (accessed October 11, 2012).
Parry, M., N. Arnell, P. Berry, D. Dodman, S. Fankhauser, C. Hope, S. Kovats, R. Nicholls, D. Satterthwaite, R. Tiffin, and T. Wheeler. 2009. *Assessing the costs of adaptation to climate change: A review of the UNFCCC and other recent estimates*. London, UK: International Institute for Environment and Development and Imperial College of London, Grantham Institute for Climate Change.
Paskal, C. 2010. The vulnerability of energy infrastructure to environmental change. *China and Eurasia Forum Quarterly* 8(2):149-163.
Patrick, S. 2007. "Failed" states and global security: Empirical questions and policy dilemmas. *International Studies Review* 9:644-662.
Patterson, K.D., and G.F. Pyle. 1991. The geography and mortality of the 1918-1919 influenza pandemic. *Bulletin of the History of Medicine* 65:4-21.
Perch-Nielsen, S.L., M. Bättig, and D. Imboden. 2008. Exploring the link between climate change and migration. *Climatic Change* 91:375-393.
Peterson, T.C., P.A. Stott, and S. Herring. 2012. Explaining extreme events of 2011 from a climate perspective. *Bulletin of the American Meteorological Society* 93:1,041-1,067.
Phillips, B.D. 2009. *Disaster recovery*. Boca Raton, FL: Taylor and Francis Group.
Phillips, B.D., D.M. Neal, and G.R. Webb. 2011. *Introduction to emergency management*. Boca Raton, FL: Taylor and Francis Group.
Pinker, S. 2011. *The better angels of our nature: Why violence has declined*. New York: Viking Adult.

Poggione, S., V.T. Gawronski, G. Hoberman, and R.S. Olson. 2012. Public response to disaster response: Applying the "5C+A" framework to El Salvador 2001 and Peru 2007. *International Studies Perspectives* 13:195–210.
Polsky, C., R. Neff, and B. Yarnal. 2007. Building comparable global change vulnerability assessments: The vulnerability scoping diagram. *Global Environmental Change* 17: 472–485.
Postel, S. 1999. *Pillar of sand: Can the irrigation miracle last?* New York: Worldwatch Institute/Norton.
Raleigh, C., and D. Kniveton. 2012. Come rain or shine: An analysis of conflict and climate variability in East Africa. *Journal of Peace Research* 49(1):51–64.
Rayner, P.J., I.G. Enting, R.J. Francey, and R. Langenfelds. 1999. Reconstructing the recent carbon cycle from atmospheric CO2, δ13C and O2/N2 observations. *Tellus* 51(2):213–232.
Remans, W. 1995. Water and war. *Hum Völkerr* 8(1):4–14.
Renaud, F.G., O. Dun, K. Warner, and J. Bogardi. 2011. A decision framework for environmentally induced migration. *International Migration* 49(Suppl. 1):e6–e29.
Reser, J.P., and J.K. Swim. 2011. Adapting to and coping with the threat and impacts of climate change. *American Psychologist* 66(4):277–289.
Reuveny, R. 2007. Climate change–induced migration and violent conflict. *Political Geography* 26(6):656–673.
Rogers, D.J., A.J. Wilson, S.I. Hay, and A.J. Graham. 2006. The global distribution of yellow fever and dengue. *Advances in Parasitology* 62:181–220.
Royal Society. 2009. *Geoengineering the climate: Science, governance and uncertainty.* London, UK: The Royal Society.
Russell, K.L., J. Rubenstein, R.L. Burke, K.G. Vest, M.C. Johns, J.L. Sanchez, W. Meyer, M. Fukuda, and D.L. Blazes. 2011. The Global Emerging Infection Surveillance and Response System (GEIS), a U.S. government tool for improved global biosurveillance: A review of 2009. *BMC Public Health* 11(Suppl. 2):S2.
Sachs, J. 2005. *The end of poverty: Economic possibilities of our time.* New York: Penguin Books.
Sachs, J. 2007. Climate change refugees. *Scientific American* 296(6):43.
Saifuddin, Y. 2012. Energy crisis: PM refrains from giving deadline. *Express Tribune*. June 28. Available: http://tribune.com.pk/story/400633/energy-crisis-pm-refrains-from-giving-deadline (accessed September 25, 2012).
Salehyan, I. 2008. From climate change to conflict? No consensus yet. *Journal of Peace Research* 45(3):315–326.
Santer, B.D., C. Mears, C. Doutriauxl P. Caldwell, P.J. Gleckler, T.M.L. Wigley, S. Solomon, N.P. Gillett, D. Ivanova, T.R. Karl, J.R. Lanzante, G.A. Meehl, P.A. Stott, K.E. Taylor, P.W. Thorne, M.F. Wehner, and F.J. Wentz. 2011. Separating signal and noise in atmospheric temperature changes: The importance of timescale. *Journal of Geophysical Research* 116:D22105.
Satterthwaite, D., S. Huq, M. Pelling, H. Reid, and P. Romero-Lankao. 2007. *Building climate change resilience in urban areas and among urban populations in low- and middle-income nations.* Center for Sustainable Urban Development. Available: http://csud.ei.columbia.edu/sitefiles/file/Final%20Papers/Week%202/Week2_Climate_IIED.pdf (accessed August 10, 2012).
Schaeffer, B., B. Mondet, and S. Touzeau. 2008. Using a climate-dependent model to predict mosquito abundance: Application to *Aedes (Stegomyia) africanus* and *Aedes (Diceromyia) furcifer* (Diptera: Culicidae). *Infection, Genetics and Evolution* 8:422–432.
Schaeffer, R., A.S. Szklo, B.S. M.C. Borba, L.P.P. Nogueira, F.P. Fleming, A. Troccoli, M. Harrison, and M.S. Boulahya. 2012. Energy sector vulnerability to climate change: A review. *Energy* 38:1–12.

Scheffran, J., M. Brzoska, J. Kominek, P.M. Link, and J. Schilling. 2012. Climate change and violent conflict. *Science* 336(6083):869–871.
Self, S. 2006. The effects and consequences of very large explosive volcanic eruptions. *Philosophical Transactions of the Royal Society A* 364:2,073–2,097.
Sen, A. 1981. *Poverty and famines*. Oxford, UK: Clarendon Press.
Sen, A. 1999. *Development as freedom*. Oxford, UK: Oxford University Press.
Seto, K.C. 2011. Exploring the dynamics of migration to mega-delta cities in Asia and Africa: Contemporary drivers and future scenarios. *Global Environmental Change* 21(Suppl. 1):S94–S107.
Seto, K.C., M. Fragkias, B. Güneralp, and M.K. Reilly. 2011. A meta-analysis of global urban land expansion. *PLoS ONE* 6(8):1–9.
Shaman, J., and M. Kohn. 2009. Absolute humidity modulates influenza survival, transmission, and seasonality. *Proceedings of the National Academy of Sciences* 106:3,243–3,248.
Shaman, J., and M. Lipsitch. 2012. The El Niño–Southern Oscillation (ENSO)–pandemic influenza connection: Coincident or causal? *Proceedings of the National Academy of Sciences*. doi: 10.1073/pnas.1107485109. Available: http://www.pnas.org/content/early/2012/01/11/1107485109.full.pdf+html (accessed August 4, 2012).
Shaman, J., V.E. Pitzer, C. Viboud, B.T. Grenfell, and M. Lipsitch. 2010. Absolute humidity and the seasonal onset of influenza in the continental United States. *PLoS Biology* 8(2):e1000316.
Shaman, J., E. Goldstein, and M. Lipsitch. 2011. Absolute humidity and pandemic versus epidemic influenza. *American Journal of Epidemiology* 173:127–135.
Siliverstovs, B., G. L'Hégaret, A. Neumann, and C. von Hirschhausen. 2005. International market integration for natural gas? A cointegration analysis of prices in Europe, North America and Japan. *Energy Economics* 27(4):603–615.
Slettebak, R.T. 2012. Don't blame the weather! Climate-related natural disasters and civil conflict. *Journal of Peace Research* 49(1):163–176.
Smith, D.J., A.S. Lapedes, J.C. de Jong, T.M. Bestebroer, G.F. Rimmelzwaan, A.D.M.E. Osterhaus, and R.A.M. Fouchier. 2004. Mapping the antigenic and genetic evolution of influenza virus. *Science* 305(5682):371–376.
Soebiyanto, R.P., F. Adimi, and R.K. Kiang. 2010. Modeling and predicting seasonal influenza transmission in warm regions using climatological parameters. *PLoS ONE* 5(3):e9450.
Solnit, R. 2010. *A paradise built in hell: The extraordinary communities that arise in disaster*. New York: Penguin Books.
Starr, J.R. 1991. Water wars. *Foreign Policy* 82:17–36.
Stein, J.G. 2010. Crisis behavior: Miscalculation, escalation, and inadvertent war. In *The international studies encyclopedia*. R.A. Denemark, Ed. Malden, MA: Wiley-Blackwell.
Stern, N. 2007. *The economics of climate change: The Stern review*. Cambridge, UK: Cambridge University Press.
Sutcliffe, J.V. 2009. The hydrology of the Nile basin. Pp. 335–364 in *The Nile: Origin, environments, limnology and human use*. H.J. Dumont, Ed. Dordrecht, The Netherlands: Springer Science+Business Media B.V.
Swart, R. 1996. Security risks of global environmental change. *Global Environmental Change* 6(3):187–192.
Tacoli, C. 2009. Crisis or adaptation? Migration and climate change in a context of high mobility. *Environment and Urbanization* 21:513–525.
Taylor, L.H., S.M. Latham, and M.E. Woolhouse. 2001. Risk factors for human disease emergence. *Philosophical Transactions of the Royal Society B* 356(1411):983–989.
Termeer, C., R. Biesbroek, and M. van den Brink. 2012. Institutions for adaptation to climate change: Comparing national adaptation strategies in Europe. *European Political Science* 11(1):41–53.

Theisen, O.M., H. Holtermann, and H. Buhaug. 2011. Climate wars? Assessing the claim that drought breeds conflict. *International Security* 36(3):79–106.
Themnér, L., and P. Wallensteen. 2012. Armed conflicts, 1946–2011. *Journal of Peace Research* 49(4):565–575.
Tietsche, S., D. Notz, J.H. Jungclaus, and J. Marotzke. 2011. Recovery mechanisms of Arctic summer sea ice. *Geophysical Research Letters* 38:L02707.
Timmer, C.P. 2010. Reflections on food crises past. *Food Policy* 35(1):1–11.
Tir, J., and D.M. Stinnett. 2012. Weathering climate change: Can institutions mitigate international water conflict? *Journal of Peace Research* 49(1):211–225.
Tol, R., and S. Wagner. 2010. Climate change and violent conflict in Europe over the last millennium. *Climate Change* 99:65–79.
Trenberth, K.E., R. Anthes, A. Belward, O. Brown, E. Haberman, T.R. Karl, S. Running, B. Ryan, M. Tanner, and B. Wielicki. 2012. Challenges of a sustained climate observing system. In *Climate science for serving society: Research, modelling and prediction priorities*. G.R. Asrar and J.W. Hurrell, Eds. Dordrecht, The Netherlands: Springer Science+Business Media B.V.
Treverton, G.F., E. Nemeth, and S. Srinivasan. 2012. *Threats without threateners? Exploring intersections of threats to the global commons and national security*. Santa Monica, CA: RAND Corporation.
United Nations. 1997. *Comprehensive assessment of the freshwater resources of the world*. Stockholm, Sweden: SEI.
United Nations. 2012. *The Millenium Development Goals report*. New York: The United Nations.
United Nations International Strategy for Disaster Reduction. 2009a. *Terminology*. Available: http://www.unisdr.org/we/inform/terminology (accessed August 10, 2012).
United Nations International Strategy for Disaster Reduction. 2009b. *2009 global assessment report on disaster risk reduction: Risk and poverty in a changing climate*. Geneva, Switzerland: United Nations.
United Nations–Water. 2008. *Transboundary water: Sharing benefits, sharing responsibilities*. Zaragoza, Spain: United Nations Office to Support the International Decade for Action "Water for Life" 2005–2015. Available: http://www.unwater.org/downloads/UNW_TRANSBOUNDARY.pdf (accessed August 10, 2012).
U.S. Climate Change Science Program and Subcommittee on Global Change Research. 2008. *Abrupt climate change*. Reston, VA: U.S. Geological Survey.
U.S. Department of Defense. 2010. *Quadrennial defense review*. Washington, DC: U.S. Department of Defense.
U.S. Department of Energy. 2005. *Department of Energy response to Hurricane Katrina*. Available: http://energy.gov/articles/department-energy-response-hurricane-katrina (accessed October 4, 2012).
U.S. Global Change Research Program. 2012. *The national global change research plan: 2012–2021*. Washington, DC: U.S. Global Change Research Program.
van Noort, S.P., R. Aguas, S. Ballesteros, and M.G. Gomes. 2011. The role of weather on the relation between influenza and influenza-like illness. *Journal of Theoretical Biology* 298:131–137.
Viboud, C., K. Pakdaman, P.-Y. Boelle, M.L. Wilson, M.F. Myers, A.-J. Valleron, and A. Flahault. 2004. Association of influenza epidemics with global climate variability. *European Journal of Epidemiology* 19:1,055–1,059.
Wamsler, C., and N. Lawson. 2012. Complementing institutional with localised strategies for climate change adaptation: A south–north comparison. *Disasters* 36(1):28–53.
Wårell, L. 2006. Market integration in the international coal industry: A cointegration approach. *Energy Journal* 27(1):99–118.

Warner, K. 2010. Global environmental change and migration: Governance challenges. *Global Environmental Change* 20(3):402–413.
Warner, K., M. Hamza, A. Oliver–Smith, F. Renaud, and A. Julca. 2010. Climate change, environmental degradation and migration. *Natural Hazards* 55:689–715.
Weaver, S.C., and W.K. Reisen. 2010. Present and future arboviral threats. *Antiviral Research* 85(2):328–345.
Weisbecker, I. (Ed.). 2011. *Climate change and human well-being: Global challenges and opportunities.* New York: Springer.
White House. 2010. *National security strategy.* Washington, DC: Office of the President. Available: http://www.whitehouse.gov/sites/default/files/rss_viewer/national_security_strategy.pdf (accessed August 10, 2012).
White House. 2012. *National strategy for global supply chain security.* Washington, DC: White House.
Wisner, B., P. Blaikie, T. Cannon, and I. Davis. 2004. *At risk: Natural hazards, people's vulnerability and disasters, 2nd ed.* London, UK: Routledge.
Witt, C.J., A.L. Richards, P.M. Masuoka, D.H. Foley, A.L. Buczak, L.A. Musila, J.H. Richardson, M.G. Colacicco-Mayhugh, L.M. Rueda, T.A. Klein, A. Anyamba, J. Small, J.A. Pavlin, M. Fukuda, J. Gaydos, K.L. Russell, and AFHSC-GEIS Predictive Surveillance Writing Group. 2011. The AFHSC-Division of GEIS Operations Predictive Surveillance Program: A multidisciplinary approach for the early detection and response to disease outbreaks. *BMC Public Health* 11(Suppl. 2):S10.
Wolf, A.T. (Ed.). 2007. Shared waters: Conflict and cooperation. *Annual Review of Environmental Resources* 32:241–269.
Wolf, A.T., J. Natharius, J. Danielson, B. Ward, and J. Pender. 1999. International river basins of the world. *International Journal of Water Resources Development* 15(4):387–427.
Wolf, A.T., S.B. Yoffe, and M. Giordano. 2003. International waters: Identifying basins at risk. *Water Policy* 5(1):29–60.
World Bank. 2005. *Pakistan country water resources assistance strategy—water economy: Running dry.* Report No. 34081-PK. Washington, DC: World Bank.
World Bank. 2011. *The World Bank supports Thailand's post-floods recovery effort.* December 13. Available: http://www.worldbank.org/en/news/2011/12/13/world-bank-supports-thailands-post-floods-recovery-effort (accessed August 10, 2012).
World Climate Research Programme. 2009. *Workshop report: World Modelling Summit for Climate Prediction.* WCRP Report No. 131. Available: http://www.wcrp-climate.org/documents/WCRP_WorldModellingSummit_Jan2009.pdf (accessed September 11, 2012).
World Economic Forum. 2012. *New models for addressing supply chain and transport risk.* Cologny, Switzerland: World Economic Forum.
Wright, B.D. 2011. The economics of grain price volatility. *Applied Economic Perspectives and Policy* 33(1):32–58.
Yancheva, G., N.R. Nowacyk, J. Mingram, P. Dulski, G. Schettler, J.F.W. Negendank, J. Liu, D.M. Seligman, L.C. Peterson, and G.H. Haug. 2007. Influence of the intertropical convergence zone on the East Asian monsoon. *Nature* 445:74–77.
Yergin, D. 2006. Ensuring energy security. *Foreign Affairs* 85(2):69–82.
Yoffe, S., A.T. Wolf, and M. Giordano. 2003. Conflict and cooperation over international freshwater resources: Indicators of basins at risk. *Journal of the American Water Resources Association* 39(5):1,109–1,126.
Zeitoun, M., and N. Mirumachi. 2008. Transboundary water interaction, I: Reconsidering conflict and cooperation. *International Environmental Agreements* 8:297–316.

Appendix A

Committee Member and Staff Biographies

John D. Steinbruner (*Chair*) is professor of public policy at the School of Public Policy at the University of Maryland and director of the Center for International and Security Studies at Maryland (CISSM). His work has focused on issues of international security and related problems of international policy. Steinbruner was director of the Foreign Policy Studies Program at the Brookings Institution from 1978 to 1996. Prior to joining Brookings, he was an associate professor in the School of Organization and Management and in the Department of Political Science at Yale University from 1976 to 1978. From 1973 to 1976 he served as associate professor of public policy at the John F. Kennedy School of Government at Harvard University, where he also was assistant director of the Program for Science and International Affairs. He was assistant professor of government at Harvard from 1969 to 1973 and assistant professor of political science at the Massachusetts Institute of Technology from 1968 to 1969. Steinbruner has authored and edited a number of books and monographs, including *The Cybernetic Theory of Decision: New Dimensions of Political Analysis* (Princeton University Press, 1974, 2002), *Principles of Global Security* (Brookings Institution Press, 2000), and *A New Concept of Cooperative Security*, co-authored with Ashton B. Carter and William J. Perry (Brookings Occasional Papers, 1992). His articles have appeared in *Arms Control Today, The Brookings Review, Daedalus, Foreign Affairs, Foreign Policy, International Security, Scientific American, Washington Quarterly,* and other journals. Steinbruner is currently co-chair of the Committee on International Security Studies of the American Academy of Arts and Sciences, chairman of the Board of the Arms Control Association, and board member of the Financial Services Volunteer Corps. He is a fellow of the American Academy of Arts and Sciences and a member of the Council on Foreign Relations. From 1981 to 2004 he was a member of the Committee on International Security and Arms Control of the National Academy of Sciences, serving as vice chair from 1996 to 2004. He was a member of the Defense Policy Board of the Department of Defense from 1993 to 1997. Born in 1941 in Denver, Colorado, Steinbruner received his A.B. from Stanford University in 1963 and his Ph.D. in political science from the Massachusetts Institute of Technology in 1968.

Otis B. Brown's specialties are Earth satellite observations, development of quantitative methods for the processing and use of satellite remotely sensed observations to study Earth system processes, and the development and application of new approaches to study climate variability and stakeholder engagement. His current research interests are observing systems, climate change impacts, adaptation strategies, and private-sector engagement. Brown served as dean of the University of Miami's Rosenstiel School of Marine and Atmospheric Science for 14 years, while being at the university for more than 40 years. He received the University of Miami Presidents Medal in honor of his outstanding leadership and distinguished accomplishments in his field of expertise as well as for his contributions to society. Brown holds a Ph.D. degree in physics, with a specialty in underwater optics, from the University of Miami; a master of science degree in theoretical physics from the University of Miami; and a bachelor of science degree in physics from North Carolina State University. Brown is a research professor at North Carolina State University.

Antonio J. Busalacchi, Jr., is the director of the Earth System Science Interdisciplinary Center (ESSIC) and a professor in the Department of Atmospheric and Oceanic Science. Busalacchi joined ESSIC in 2000 after serving as chief of the NASA/Goddard Laboratory for Hydrospheric Processes. He has studied tropical ocean circulation and its role in the coupled climate system. His interests include the study of climate variability and prediction, tropical ocean modeling, ocean remote sensing, and data assimilation. His research in these areas has supported a range of international and national research programs dealing with global change and climate, particularly as affected by the oceans. From 1989 to 1996 he served on the National Academy of Sciences/National Research Council (NAS/NRC) Tropical Ocean Global Atmosphere Advisory Panel and from 1991 to 1993 he was a member of the NAS/NRC Panel on Ocean Atmosphere Observations Supporting Short-Term Climate Predictions. From 1999 to 2006 he served as cochairman of the Scientific Steering Group for the World Climate Research Programme on Climate Variability and Predictability (CLIVAR). From

2003 to 2008 he served as chairman of the NAS/NRC Climate Research Committee and from 2007 to 2008 as chair of the NAS/NRC Committee on Earth Science and Application: Ensuring the Climate Measurements from the National Polar-Orbiting Operational Environmental Satellite and the Geostationary Operational Environmental Satellite-R Series Programs. Presently, he serves as chair of the Joint Scientific Committee for the World Climate Research Programme and chair of the NAS/NRC Board on Atmospheric Sciences and Climate. He is a fellow of the American Meteorological Society (AMS), American Geophysical Union, and in 2006 he was selected by the AMS to be the Walter Orr Roberts Interdisciplinary Science Lecturer. He received his Ph.D. in oceanography from Florida State University in 1982.

David Easterling is currently chief of the Scientific Services Division at National Oceanic and Atmospheric Administration's National Climatic Data Center in Asheville, North Carolina. He served as an assistant professor in the Climate and Meteorology Program, Department of Geography, Indiana University–Bloomington from 1987 to 1990. In 1990 he moved to the National Climatic Data Center as a research scientist, was appointed principal scientist in 1999, and chief of scientific services in 2002. He has authored or co-authored more than 60 research articles in journals such as *Science*, *Nature*, and the *Journal of Climate*. Easterling was also a contributor to the Intergovernmental Panel on Climate Change (IPCC) Second and Third Assessment Reports, and a lead author for the IPCC Fourth Assessment Report. He was a convening lead author for the U.S. Climate Change Science Plan Synthesis and Assessment Product on Climate Extremes and is a lead author of the chapter on the natural physical environment of the IPCC Special Report on Extreme Events. His research interests include the detection of climate change in the observed record, particularly changes in extreme climate events. He received his Ph.D. from the University of North Carolina at Chapel Hill in 1987.

Kristie L. Ebi is a consulting professor in the Department of Medicine at Stanford University and an independent consultant. She conducts research on the impacts of and adaptation to climate change, including on extreme events, thermal stress, food-borne safety and security, and vector-borne diseases. Her work focuses on understanding sources of vulnerability and designing adaptation policies and measures to reduce the health risks of climate change in a multi-stressor environment, including identifying indicators to measure changes in resilience and effectiveness of adaptation options. She has worked with the World Health Organization, the United Nations Development Programme, the U.S. Agency for International Development, and others on assessing vulnerability and implementing adaptation measures in Central America, Europe, Africa, Asia, and the Pacific. She facilitated adaptation assessments for the health sector for the states of Maryland and Alaska. She was a coordinating lead author or lead author for the human health assessment for SAP4.6, the first U.S. National Assessment, the IPCC Fourth Assessment Report, the Millennium Ecosystem Assessment, and the International Assessment of Agricultural Science and Technology for Development. Ebi's scientific training includes an M.S. in toxicology and a Ph.D. and a master of public health degree in epidemiology, and two years of postgraduate research at the London School of Hygiene and Tropical Medicine. She has edited four books on aspects of climate change and has more than 100 publications.

Thomas Fingar is the Oksenberg-Rohlen Distinguished Fellow and a senior scholar at the Freeman Spogli Institute for International Studies at Stanford University. From May 2005 through December 2008 he served as the first deputy director of National Intelligence for Analysis and, concurrently, as chairman of the National Intelligence Council. Fingar served previously as assistant secretary of the State Department's Bureau of Intelligence and Research (2004–2005), principal deputy assistant secretary (2001–2003), deputy assistant secretary for analysis (1994–2000), director of the Office of Analysis for East Asia and the Pacific (1989–1994), and chief of the China Division (1986–1989). Between 1975 and 1986 he held a number of positions at Stanford University, including senior research associate in the Center for International Security and Arms Control. Fingar is a graduate of Cornell University (A.B. in government and history, 1968) and Stanford University (M.A., 1969, and Ph.D., 1977, both in political science).

Leon Fuerth is the former national security adviser to Vice President Al Gore and the founding director of the Project on Forward Engagement. As the Vice President's national security advisor, he served on the Principals' Committee of the National Security Council alongside the secretary of state, the secretary of defense, and the President's own national security adviser. Fuerth organized and managed five bi-national commissions with Russia, South Africa, Egypt, Ukraine, and Kazakhstan. Before beginning his work on Capitol Hill in 1979, he spent 11 years as a foreign service officer, serving in such places as the U.S. consulate in Zagreb and the state department. In 2001 Fuerth founded the Project on Forward Engagement to explore methods for incorporating systematic foresight into the policy process. The project focuses on developing "anticipatory governance," a system of systems to (a) integrate foresight and policy, (b) network across governance, and (c) rapidly apply learning to policy and operations. The project is based out of the Elliott School of International Affairs at the George Washington University, where Fuerth holds an appointment as a research professor, and also operates at the National Defense University, where he is appointed as a distinguished research fellow. He holds a bachelor's degree in English and a master's degree in history from New York University, as well as a master's degree in public administration from Harvard University.

Sherri Goodman is senior vice president, general counsel, and corporate secretary of CNA and serves as executive director of CNA's Military Advisory Board. Goodman is an internationally recognized authority on energy, climate change, and national security, having led the projects by CNA's Military Advisory Board on National Security and the Threat of Climate Change (2007) and Powering America's Defense: Energy and the Risks to National Security. From 1993 to 2001 Goodman was deputy undersecretary of defense (environmental security), serving as the chief environmental, safety, and occupational health officer for the Department of Defense. In this position she was responsible for more than $5 billion in annual defense spending, including programs on energy efficiency and climate change, cleanup at active and closing bases, compliance with environmental laws, environmental cooperation with foreign militaries, and conservation of natural and cultural resources. Goodman received a J.D. cum laude from the Harvard Law School and a master's degree in public policy from Harvard's John F. Kennedy School of Government. She received her B.A. summa cum laude from Amherst College.

Jo L. Husbands is a scholar and senior project director with the Board on Life Sciences of the U.S. National Academy of Sciences (NAS), where she manages studies and projects to help mitigate the risks of the misuse of scientific research for biological weapons or bioterrorism. She also represents the NAS on the Biosecurity Working Group of the IAP, the global network of science academies, which also includes the academies of China, Cuba, Nigeria, Poland, and the United Kingdom. From 1991 to 2005 she was director of the NAS Committee on International Security and Arms Control (CISAC) and its Working Group on Biological Weapons Control. Before joining the National Academies, she worked for several Washington, DC-based nongovernmental organizations focused on international security. Husbands is currently an adjunct professor in the Security Studies Program at Georgetown University, where she teaches a course on the international arms trade. She holds a Ph.D. in political science from the University of Minnesota and a master's degree in international public policy (international economics) from the Johns Hopkins University School of Advanced International Studies.

Robin Leichenko is an associate professor in the Department of Geography at Rutgers University. Her research addresses the urban and regional impacts of global economic and environmental change in both advanced and developing countries. She recently co-authored *Environmental Change and Globalization: Double Exposures*, which focuses on how processes of globalization and climate change jointly affect vulnerable regions, social groups, and ecosystems. Other current research includes a study of the effects of the globalization of consumption practices on housing demand and suburbanization patterns in China and the United States, a study of climate change vulnerability and adaptation in U.S. cities, and a study of the effects of globalization trends on U.S. firms and workers. Recently completed research projects include a study of the impacts of international trade on employment and income inequality across U.S. regions and a study of the effects of globalization and climate change on rural agricultural regions in India and Southern Africa. She earned her Ph.D. in geography from Penn State.

Robert J. Lempert is director of the Frederick S. Pardee Center for Longer Range Global Policy and the Future Human Condition at the RAND Corporation. He was a member of the 2007 Nobel Peace Prize–winning Intergovernmental Panel on Climate Change. Lempert is an internationally known scholar in the field of decision making under conditions of deep uncertainty. He is a member of the Council on Foreign Relations, a fellow of the American Physical Society, and a member of the National Academy of Science's Climate Research Committee. His research focuses on improving methods for long-term policy analysis and for using data and models to support decision making where accurate forecasts are impossible. He is leading a major National Science Foundation–funded study that aims to improve methods for using scientific and other information to support decisions about climate change. He has worked extensively in the areas of environment, energy, and national security strategies; and he has conducted research on science and technology investment strategies for clients that include the White House Office of Science and Technology Policy, the U.S. Department of Energy, the National Science Foundation, and several multinational firms.

Marc Levy is deputy director of the Center for International Earth Science Information Network (CIESIN), a unit of Columbia University's Earth Institute. He is also an adjunct professor in Columbia's School of International and Public Affairs. He is a political scientist specializing in the human dimensions of global environmental change. His research focuses on climate–security linkages, emerging infectious disease modeling, anthropogenic drivers of global change, sustainability indicators, and vulnerability mapping. He is also leading a project in Haiti to reduce vulnerability to disaster risks by integrating ecology and economic development goals on a watershed scale. He has served on a number of international assessments, and is currently a lead author on the Intergovernmental Panel for Climate Change Fifth Assessment Report chapter on human security.

David Lobell is an assistant professor at Stanford University in environmental earth system science and an associate director in Stanford's Center on Food Security and the Environment. His research focuses on identifying opportunities to raise crop yields in major agricultural regions, with a particular emphasis on adaptation to climate change. He is a fellow of the American Geophysical Union and received the 2010 James B. Macelwane Medal. He is currently serving as lead author on the "Food Production Systems and Food Security" chapter of the Intergovernmental Panel on Climate Change (IPCC) Fifth Assessment Report. Lobell received a Ph.D. in geological and environmental sciences from Stanford University in 2005 and a Sc.B. in applied mathematics from Brown University in 2000.

Richard Stuart Olson is director of extreme event research and professor in the Department of Politics and International Relations at Florida International University. A Fulbright Fellow in Colombia in 1968–1969, he returned to Latin America in 1972 to conduct field research on the Managua, Nicaragua, earthquake disaster of that year. Since then he has been directly involved in disaster response, evaluation, and research in more than 20 events, including Guatemala in 1976 (earthquake); Chile in 1985 (earthquake); Mexico City in 1985 (earthquakes); Colombia in 1985 (volcanic eruption and lahar) and 1994 (earthquake and landslide); the Dominican Republic, Honduras, and Nicaragua in 1998 (hurricanes); and El Salvador in 1986 and 2001 (earthquakes). In addition to more than 60 research articles, monographs, and major papers, Olson was lead author on the books *The Politics of Earthquake Prediction* (Princeton University Press, 1989) and *Some Buildings Just Can't Dance: Politics, Life Safety, and Disaster* (Elsevier/JAI, 1999). He received a B.A. from the University of California, Davis, in 1967; an M.A. from the University of California, Los Angeles, in 1968; and a Ph.D. in 1974 from the University of Oregon, all in political science and emphasizing comparative and Latin American politics.

Richard L. Smith is Mark L. Reed III Distinguished Professor of Statistics and professor of biostatistics at the University of North Carolina, Chapel Hill, and director of the Statistical and Applied Mathematical Sciences Institute. His expertise is in statistical aspects of climate change research and air pollution health effects. Smith is a fellow of the American Statistical Association and the Institute of Mathematical Statistics and an elected member of the International Statistical Institute, and he won the Guy Medal in Silver of the Royal Statistical Society and the Distinguished Achievement

Medal of the Section on Statistics and the Environment from the American Statistical Association. In 2004 he was the J. Stuart Hunter Lecturer of The International Environmetrics Society (TIES). He is also a chartered statistician of the Royal Statistical Society. He obtained his Ph.D. from Cornell University.

Paul C. Stern is a senior scholar at the National Research Council/National Academy of Sciences, working primarily with the Board on Environmental Change and Society, formerly known as the Committee on Human Dimensions and Global Change. His work at the National Research Council has included directing studies on climate and global change, such as *Informing Decisions in a Changing Climate* (2009), *Decision Making for the Environment: Social and Behavioral Science Priorities* (2005), and *Global Environmental Change: Understanding the Human Dimensions* (1992), and he has been involved in the suite of *America's Climate Choices* studies. His work has also included studies on international security issues that have produced reports such as *International Conflict Resolution After the Cold War* (2000) and a three-volume series on *Behavior, Society, and International Conflict* (1989–1993). His research interests include the determinants of environmentally significant behavior, particularly at the individual level; participatory processes for informing environmental decision making; processes for informing environmental decisions; and the governance of environmental resources and risks. He is coauthor of the textbook *Environmental Problems and Human Behavior* (2nd ed., 2002) and of the 2003 article "The Struggle to Govern the Commons," which won the 2005 Sustainability Science Award from the Ecological Society of America. He is a fellow of the American Association for the Advancement of Science and of the American Psychological Association. He holds a B.A. from Amherst College and an M.A. and Ph.D. from Clark University, all in psychology.

Appendix B

Briefings Received by the Committee

January 12, 2012. Daniel Cooley, Colorado State University. Estimating probabilities of climate events in the joint tail.
January 12, 2012. David R. Easterling, National Oceanic and Atmospheric Administration (NOAA). Summer temperatures and drought.
January 12, 2012. Lisa Goddard, Columbia University. Looking to the future: Decadal variability and its prediction in dynamical models.
January 12, 2012. Mary Hayden, National Center for Atmospheric Research. The dengue vector mosquito *Aedes aegypti* at the margins: Sensitivity of a coupled natural and human system to climate change.
January 12, 2012. Upmanu Lall, Columbia University. Floods: National Research Council study climate change and security.
January 12, 2012. Robert Lempert, RAND Corporation. Intergovernmental Panel on Climate Change (IPCC) Special Report on Managing the Risks of Extreme Events and Disasters to Advance Climate Change Adaptation.
January 12, 2012. Richard L. Smith, University of North Carolina and Statistical and Applied Mathematical Sciences Institute. Estimating probability of an extreme weather event in the next ten years.
January 12, 2012. Gabriel Vecchi, NOAA and Geophysical Fluid Dynamics Laboratory. Cat. 4–5 landfalling tropical cyclones in the coming decade.

January 13, 2012. Kristie L. Ebi, Carnegie Institution of Washington. IPCC Special Report on Managing the Risks of Extreme Events and Disasters to Advance Climate Change Adaptation.

March 1, 2012. Daniel P. Aldrich, Purdue University and U.S. Agency for International Development (USAID). The past, present, and future of the Fukushima nuclear crisis.

March 1, 2012. Marc F. Bellemare, Duke University. Rising food prices, food price volatility, and social unrest.

March 1, 2012. Gary Eilerts, USAID and Chris Funk, U.S. Geological Survey. Climate and related factors that influence African food security: African food security conditions are an important feature of U.S. national security.

March 1, 2012. Thomas M. Parris, ISciences. Environmental indications and warnings.

March 1, 2012. Kaitlin Shilling, Stanford University. Climate and conflict.

March 1, 2012. Mark L. Wilson, University of Michigan. Climate impacts on emerging and re-emerging diseases.

March 2, 2012. Neil Adger, University of East Anglia. Migration, climate change, and national security.

March 2, 2012. Jack A. Goldstone, George Mason University. Forecasting political instability: Political Instability Task Force history and approach.

March 2, 2012. Joseph Hewitt, University of Maryland. 2011 alert lists: Methodological overview.

March 2, 2012. Sari Kovats, London School of Hygiene and Tropical Medicine. Climate change and health.

March 2, 2012. Jürgen Scheffran, Universität Hamburg. Conflict sensitivity in climate hot spots.

Appendix C

Method for Developing Figure 3-1

To develop Figure 3-1 we applied the methods used in Easterling and Wehner (2009) to examine the chances that any 10-year period in the globally averaged annual temperature time series will have a negative or positive trend. Here we extended this method two ways. First, in addition to the global analysis we did the same analysis for each region shown in Figure 3-1. Second, we used projected surface temperatures from the latest state-of-the-art climate models that are being used in the development of the Intergovernmental Panel on Climate Change (IPCC) Fifth Assessment Report (AR5) and available from the Coupled Model Intercomparison Project Phase 5 (CMIP5) database (Taylor et al., 2012)

We used annual averaged surface temperature projections (e.g., one temperature value per year) from simulations of the 21st century by six different global climate modeling groups from the publicly available CMIP5 database. The models include the Canadian Earth System Model (CANESM2), the National Center for Atmospheric Research Community Climate System Model (CCSM4), the French National Center for Meteorological Research climate model (CNRM-CM5), the National Oceanic and Atmospheric Administration Geophysical Fluid Dynamics Laboratory (GFDL) climate model (GFDL-CM3), two GFDL earth system models (GFDL-ESM2G and GFDL-ESM2M), and the Norwegian Earth System Model. The models were run using the Representative Concentration Pathway (RCP) 8.5 (Moss et al., 2010) greenhouse gas (GHG) forcing scenario developed for the IPCC AR5 and available from the Earth System Grid portal (http://www.earthsystemgrid.org [accessed November 15, 2012]). Each modeling group provided a set or ensemble of two climate simulations for the 21st century each forced with the RCP8.5 scenario but starting with slightly different initial conditions to represent how the climate might evolve for the 2000–2099 period. The RCP8.5 scenario is close to the SRES A2 scenario used in previous IPCC reports and represents a business as usual GHG increase of about 1 percent per year. Discussions of climate models and their limitations are beyond the scope of this document, but some can be found in Chapter 8 of the IPCC Fourth Assessment Report by Working Group I (Randall et al., 2007).

An annually averaged surface temperature from each ensemble member for the 2000–2050 period was used to calculate an annual temperature time series for each region in Figure 3-1 and for the globe by averaging the temperature across all model grid points in each region or for the globe, which resulted in one time series of temperature for each region for each model simulation (see Figure 2 from Easterling and Wehner, 2009, for an example). This resulted in 14 time series for each region (seven models, two simulations for each model). Ordinary least squares (OLS) trends were then calculated for all running 10-year periods (e.g., 2000–2009, 2001–2010, 2002–2011, etc.) for each region's time series for each model simulation. Each region had 588 trends calculated because there are 42 overlapping 10-year periods for each time series and 14 total time series for each region. All trends for a given region were used to construct each probability distribution function shown in Figure 3-1.

We restricted our analysis to the 2000–2050 model period because the simulated change in global air temperature for this period generally is linear, with acceleration typically starting after 2050. Additionally, the time evolution of simulated global surface air temperature for the 2000–2050 period differs little between the different RCP forcing scenarios, with differences becoming clear only after 2050. This allows us to generalize the results shown in Figure 3-1 to any given 10-year period between 2000 and 2050 across a wide range of GHG emissions trajectories in the coming decades.

REFERENCES

Easterling, D.R., and M.F. Wehner. 2009. Is the climate warming or cooling? *Geophysical Research Letters* 36:8. doi:10.1029/2009GL037810.

Moss, R.H., J.A. Edmonds, K.A. Hibbard, M.R. Manning, S.K. Rose, D.P. van Vuuren, T.R. Carter, S. Emori, M. Kainuma, T. Kram, G.A. Meehl, J.F.B. Mitchell, N. Nakicenovic, K. Riahi, S.J. Smith, R.J. Stouffer, A.M. Thomson, J.P. Weyant, and T.J. Wilbanks. 2010. The next generation of scenarios for climate change research and assessment. *Nature* 463:747–756.

Randall, D.A., R.A. Wood, S. Bony, R. Colman, T. Fichefet, J. Fyfe, V. Kattsov, A. Pitman, J. Shukla, J. Srinivasan, R.J. Stouffer, A. Sumi, and K.E. Taylor. 2007. Climate models and their evaluation. Pp. 589–662 in *Climate change 2007: The physical basis. Contribution of Working Group I to the Fourth Assessment Report of the Intergovernmental Panel on Climate Change*, S. Solomon, D. Qin, M. Manning, Z. Chen, M. Marquis, K.B. Averyt, M. Tignor, and H.L. Miller, Eds. New York: Cambridge University Press.

Taylor, K.E., R.J. Stouffer, and G.A. Meehl. 2012. An overview of CMIP5 and the experiment design. *Bulletin of the American Meteorological Society* 93:485–498.

Appendix D

Statistical Methods for Assessing Probabilities of Extreme Events

Although it is difficult to predict individual extreme events on a decadal time-scale, there are nevertheless studies of how probabilities of extreme events have changed over the time period of the observed record as well as projections of possible future changes using climate models. To the best of our knowledge, such methods have not been used to project probabilities of extreme events over the decadal time-scale that is the specific focus of this report. We outline here some of the relevant concepts because in our view it would be fully feasible to obtain such projections.

METHODS BASED ON TIME SERIES ANALYSIS

In time series analysis, a sequence of observations (e.g., annual temperature means over a specified region) are analyzed statistically for changes in their means and variances. In most realistic settings, the observations are not statistically independent, so account must be taken of autocorrelation among consecutive observations. Standard methods for doing this have been presented in numerous texts, e.g., Brockwell and Davis (2002).

As an example, Hansen et al. (2012) provided empirical analyses of temperature means from numerous parts of the world and noted the increasing frequency of extreme events, which they defined as events that are more than three standard deviations from the 1951–1980 mean. An analysis of this nature is useful for documenting the increasing frequency of extreme events but does not lead to quantitative projections for the future.

A more statistically sophisticated approach is represented by Rahmsdorf and Coumou (2011). Treating observations as independent and Gaussian distributed, but with changing means, they provided concrete formulas for calculating the probability that a specific threshold is exceeded over a given period of time. However, the restriction to independent observations limits the usefulness of their approach.

An alternative is to use Monte Carlo simulation. Based on any fitted time series model, such as an autoregressive moving average (ARMA) model with time-varying means, one could calculate future extreme event probabilities by simulating the time series many times and calculating the proportion of simulations for which the extreme event of interest occurs. The principal limitation of such methods lies not in the simulation itself, which is fast and accurate using modern computing techniques, but in the structure of the time series model; if this is misspecified, then the extreme value probabilities may be over- or under-estimated by orders of magnitude.

In particular, we question whether Gaussian probabilities are appropriate for extreme events. In the case of temperature series, simple plots of the data do show an approximately normal shape (Hansen et al., 2012, have several examples), but this does not preclude the possibility of some extreme events that are caused by natural variation in the weather. An example is given by Dole et al. (2011), where they dispute the assertion that the 2010 Russian heat wave was associated with anthropogenic climate change. A key point of their argument was the presence of a blocking event, which could be a natural occurrence, but one that is expected to lead to much more extreme temperatures than usual. In the presence of such phenomena, one would not expect the distributions of the most extreme events to be consistent with Gaussian probabilities. For other kinds of meteorological variables, such as precipitation and wind speed, it is not realistic to assume Gaussian probabilities at all.

METHODS BASED ON EXTREME VALUE THEORY

An alternative class of statistical methods, which do not assume Gaussian probabilities, is the class of methods based on the statistical theory of extreme values. One very readable book on this topic is by Coles (2001). According to classical extreme value theory, the distribution of maxima over a fixed time period (e.g., annual maximum temperatures) may be approximated by a family of probability distributions known as extreme value distributions. These may be summarized in the form of the Generalized Extreme Value (GEV) distribution, which is characterized by three parameters: a location parameter representing the center of the distribution, a scale parameter representing variability, and a third "shape" parameter, which is the key parameter in characterizing probabilities of very extreme events. Although in their original form such methods were developed for stationary (but not necessarily Gaussian) time series, they are easily adapted to nonstationary time series by allowing the three GEV parameters to be time varying.

An extension of classical extreme value theory is to use threshold methods. Rather than characterize extreme events solely in terms of the distribution of annual maxima, threshold methods take account of all events beyond a given high threshold, also known as exceedances. A common probability distribution for events beyond a threshold is the Generalized Pareto distribution, which has properties very similar to those of the GEV distribution. In this case, also, one can take account of time dependence by allowing the parameters of the model to vary with time, thus creating a theoretical framework for calculating the probabilities of extreme events in the presence of a changing climate.

Kharin and Zwiers (2000) were possibly the first to use these methods to explore the effect of climate change on extremes. Recent contributions include assessing the role of anthropogenic influence on extremes of temperature (Zwiers et al., 2011) and precipitation (Min et al., 2011). Wehner et al. (2010) used the GEV distribution to compare observational and model-based calculations of precipitation extremes, showing that the agreement is highly dependent on the resolution of the climate model.

For all of these methods, the emphasis in most climate studies has been on relatively long time horizons (e.g., 40 years), but the same statistical models can be used (via simulation) to estimate probabilities over shorter time horizons, such as 10 years.

ATTRIBUTION OF INDIVIDUAL EXTREME EVENTS

As noted elsewhere in this report, our primary focus has not been on the question of attribution, that is, on whether observed climate change is due to anthropogenic factors as opposed to natural forcing or internal variability. Nevertheless, some of the analytic methods developed in that context are relevant to understanding how extreme event probabilities might change over a time span of 10 years or less.

In a paper motivated by the 2003 heat wave in Europe, Stott et al. (2004) calculated summer (June, July, August) annual temperature averages over a large area of western Europe and used climate models both with and without anthropogenic forcing to estimate the probability of an extreme event under either scenario. Their statistical methodology used a conventional detection and attribution approach to decompose the observational time series into components due to anthropogenic forcing, natural forcing, and internal variability, combined with the Generalized Pareto distribution, fitted to events beyond a high threshold, to estimate probabilities of extreme events. More recently, Pall et al. (2011) showed how to extend the methodology to much smaller temporal and spatial scales using large numbers of climate model runs focused on the temporal/spatial scale of interest. Leaving aside the attribution question, these papers provide two specific examples of how observational and climate model data may be combined to estimate the probability of an observed extreme event under current climate conditions and, by extension, how those probabilities might change under scenarios of future climate change.

EXTREMES OF DEPENDENT EVENTS (EVENT CLUSTERS)

We are also concerned with the possibility of extreme events occurring simultaneously in different locations as a result of common meteorological features such as ENSO or Rossby waves. Statistical methods have been developed for this problem, primarily through the methods of multivariate extreme value theory. The bivariate case (where there are just two dependent events) has been particularly highly studied.

In its simplest form, the method used for bivariate analysis is first to perform an analysis of the two variables individually (either a GEV or Generalized Pareto analysis could be appropriate for this) and then to transform the distribution to unit Fréchet form [$P(X<x)=e^{-1/x}$, $x>0$] using a probability integral transform. This transformation has the effect of exaggerating the most extreme events so that they stand out sharply on a plot. Traditional measures of dependence, such as correlation, are not readily interpretable in this context, but a number of alternative measures of dependence that are specifically adapted to extreme events have been proposed (Coles et al., 1999).

Going beyond simple characterizations of extremal dependence, there are a number of formal statistical models that have been used to calculate joint probabilities of extreme events. There has been limited practical application of these models to climate data, but we illustrate the possibilities by considering two examples related to earlier discussion.

Example 1. Herweijer and Seager (2008) argued that the persistence of drought patterns in various parts of the world may be explained in terms of sea surface temperature patterns. One of their examples (Figure 3 of their paper) demonstrated that precipitation patterns in the south-west United States are highly correlated with those of a region of South America including parts of Uruguay and Argentina. As an illustration of this, we have computed annual precipitation means corresponding to the same regions that they defined, and we show a scatterplot of the data in the left-hand panel of Figure D-1. The two variables are clearly correlated (r = 0.38; p <.0001). The correlation coefficient is lower than that found by Herweijer and Seager (r = 0.57), but this is explained by their use of a six-year moving average filter, which naturally increases the correlation. However, the feature of interest to us here is not the correlation in the middle of the

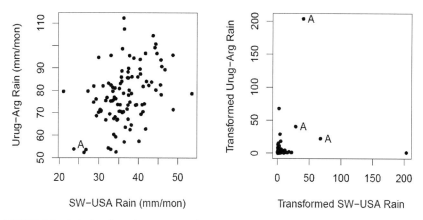

FIGURE D-1 Left: Plot of U.S. annual precipitation means over latitudes 25–35°N, longitudes 95–120°W, against Argentina annual precipitation means over latitudes 30–40°S, longitudes 50–65°W, 1901–2002. Right: Same data with empirical transformation to unit Fréchet distribution. Observations near the letter A in the left-hand plot and marked by A in the right-hand plot refer to simultaneous occurrences of extremely low precipitation in both locations. Data from gridded monthly precipitation means archived by the Climate Research Unit of the University of East Anglia (http://www.cru.uea.ac.uk/cru/data/hrg/timm/grid/CRU_TS_2_1.html [accessed November 15, 2012]).

distribution, but instead the dependence that exists in the lower tail (lower tail rather than upper tail, because our focus is drought). Therefore, the variables are transformed empirically to the unit Fréchet distribution (small values of precipitation corresponding to large values on Fréchet scale), with the results shown in the right-hand panel of Figure D-1. (Supplementary information on the methods and programs used in our analyses is available at http://sites.nationalacademies.org/DBASSE/BECS/DBASSE_073118 [accessed November 15, 2012].)

The effect of the Fréchet transformation is to highlight the most extreme observations in each variable. However, the most interesting observations are those that are not close to either of the axes, because these correspond to observations that are extreme in both variables. In particular, the triangle of observations near the letter A in the left-hand plot are transformed into the observations marked A in the right-hand plot, which are all far from either axis. This is empirical evidence that there is indeed dependence between the most extreme values in this example.

To go further, we have fitted one of the standard extremal dependence models—the logistic model, for which a detailed methodology based on events exceeding a threshold was developed by Coles and Tawn (1991), although the model itself goes back to Gumbel and Mustafi (1967). We have used rather a low threshold (2.5 on the unit Fréchet scale) in order to illustrate the applicability of the method; ideally, we would like to use a longer series and a higher threshold. An intuitive way to understand the effect of this model is to show how the probability of a jointly extreme event in both variables is inflated compared with what it would be if the variables were independent. For example, if we consider the 10-year return level (the value of each variable that would be exceeded with a probability of 1/10 in a single year), if the variables were independent, the probability that both 10-year return values would be exceeded is $(1/10)^2$ or 0.01. Under the logistic model fitted to this dataset, the joint probability is 0.027—an increase of 2.7 over the independent case. For more extreme events, the relative increase in joint probability compared with the independent case is larger—4.7 for the 20-year return level, and 10.8 for the 50-year return level. However, confidence intervals for these relative increases in joint probability are quite wide. For example, for the 50-year return level, a 90 percent confidence interval is (2.1, 18.8), obtained by bootstrapping.

The logistic model, although very widely used in bivariate extreme value modeling, has a couple of well-documented disadvantages: It assumes symmetry between the two variables, and it has also a property known as asymptotic dependence, which might not be satisfied in practice. Recent work by Ramos and Ledford (2009, 2011) has suggested an alternative, more complicated, model that does not make those assumptions. They called this model the η-asymmetric logistic model, but for the present discussion we shall call it the Ramos–Ledford model. The estimation procedure used here follows Section 4.1 of Ramos and Ledford (2009). Under this model, the estimated probability ratios are very similar to those of the logistic model, although the confidence intervals are somewhat wider. A summary of all the estimates and confidence intervals is in Table D-1.

TABLE D-1 Estimates of the Increase in Probability of a Joint Extreme Event in Both Variables, Relative to the Probability Under Independence, for the United States/Uruguay–Argentina Precipitation Data

	Logistic Model		Ramos-Ledford Model	
	Estimate	90% CI	Estimate	90% CI
10-year	2.7	(1.2, 4.2)	2.9	(1.2, 5.0)
20-year	4.7	(1.4, 7.8)	4.9	(1.2, 9.6)
50-year	10.8	(2.1, 18.8)	9.9	(1.4, 23.4)

NOTE: Shown are the point estimate and 90 percent confidence interval, under both the logistic model and the Ramos-Ledford model.

Example 2. Lau and Kim (2012) have provided evidence that the 2010 Russian heat wave and the 2010 Pakistan floods were derived from a common set of meteorological conditions, implying a physical dependence between these two very extreme events. Using the same data source as for Example 1, we have constructed summer temperature means over Russia and precipitation means over Pakistan corresponding to the spatial areas used by Lau and Kim. Figure D-2 shows a scatterplot; the left-hand plot is of the raw data, and the right-hand plot is of the data after transformation to the unit Fréchet distribution (with the largest values on the original plot corresponding to the largest value on Fréchet scale, because the right-hand tail is of interest here). Because the data source goes up only to 2002, we have approximated the 2010 values using a different data source (the National Centers for Environmental Prediction); this data point is shown in the left-hand panel of Figure D-2 but is not included in the subsequent analysis. The 2010 value is clearly an outlier for temperature but not for precipitation. It should be noted that, while the 2010 Pakistan flooding was severe, the overall rainfall over northern Pakistan was not unprecedented. This is because the heavy rain was concentrated in a very small area over the upper Indus river basin, over a few days (Dr. W.K. Lau, Chief of Atmospheres, National Aeronautics and Space Administration, 2012, personal communication).

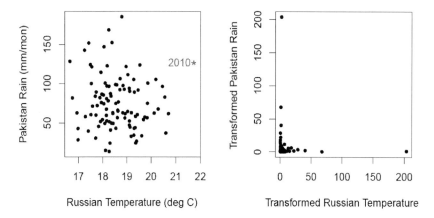

FIGURE D-2 Left: Plot of June, July, and August (JJA) Russian temperature means against Pakistan JJA precipitation means, 1901–2002. Right: Same data with empirical transformation to unit Fréchet distribution. Data from Climatic Research Unit, as in Figure D-1. The Russian data were averaged over 45–65°N, 30–60°E, while the Pakistan data were averaged over 32–35°N, 70–73°E, same as in Lau and Kim (2012).

TABLE D-2 Similar to Table D-1, but for the Russia-Pakistan Dataset

	Logistic Model		Ramos-Ledford Model	
	Estimate	90% CI	Estimate	90% CI
10-year	1.01	(1.00, 1.01)	0.33	(0.04, 1.4)
20-year	1.02	(1.00, 1.03)	0.21	(0.008, 1.8)
50-year	1.05	(1.01, 1.07)	0.17	(0.001, 2.9)

NOTE: Shown are the point estimate and 90 percent confidence interval, under both the logistic model and the Ramos-Ledford model.

In contrast with Figure D-1, the right hand plot of Figure D-2 shows virtually no data point away from the axes, indicating that there is no evidence of dependence in the upper tail of the distribution. This is confirmed by repeating the same analyses as for Example 1, with results shown in Table D-2. For the logistic model, which is constrained to positive dependence between the two variables, the point estimates and confidence intervals (for the ratio of joint probability to the independent case) are all very close to 1. Under the Ramos-Ledford model, which does not have that constraint, the estimated probability ratios are <1 (indicating negative dependence), but the confidence intervals include 1. With either set of results, the net conclusion is that there is no evidence against the hypothesis of independence in the right hand tail of the distribution.

Conclusions. Example 1 confirms and extends the results of Herweijer and Seager (2008) by showing that the interdependence of drought conditions in the two given regions of the United States and South America extends to the tail of the distribution, although the confidence intervals for the probability ratios are still fairly wide as a result of the relatively small number of data points (102). However, Example 2 shows no evidence at all that there is any tendency for extreme high temperatures in Russia to be associated with extreme high precipitation in Pakistan; in other words, the 2010 event may have been truly an outlier without precedent in history. This should however be qualified by noting that the dataset used, consisting of monthly averages over half-degree grid cells, cannot be expected to reproduce extreme precipitation events over very short time and spatial scales, and it remains possible that an alternative data source, using finer-scale data, would produce a different conclusion.

FUTURE RESEARCH NEEDS

There is a substantial body of statistical literature on univariate and bivariate extremes and more limited research on extremes in higher dimensions. However, practical application of these methods in extreme value statistics to the kinds of problems considered in the present report has been limited. For the univariate case, methods exist for determining trends in extreme event probabilities based on observational data and, by combining observations with climate models, for extrapolating these trends forward in time. However, we are not aware of published results that directly address the question of a 10-year time frame, which has been the main focus of the present report. There is no reason in principle that existing statistical methods could not be used to produce such estimates, and we recommend pursuing that.

For the bivariate case, the main question of interest is one of dependence: whether some underlying process creates a reliable association between the occurrence of an extreme event in one climate variable in one place and the probability of an extreme event in another variable or place. The relatively short length of most observational series limits the extent to which this question can be answered based on observational data. It would be valuable to conduct studies using longer series generated from climate models. Another question concerns the time-scale of dependence, e.g., if one extreme does indeed raise the probability of another, then for what period of time does this elevated probability of an extreme event remain valid?

More broadly, there is a need for research on clusters of extreme events. It is possible that an extreme value of one climate variable is systematically associated with extreme events in several related variables. Extensions of extreme value theory to multivariate data, to time series and spatial processes (see, e.g., Cooley et al., 2012), could in principle be used to answer such questions, but there is a need for more extensive practical development of these methods.

REFERENCES

Brockwell, P.J., and Davis, R.A. 2002. *Introduction to time series and forecasting* (2nd edition). New York: Springer Verlag.
Coles, S.G. 2001. *An introduction to statistical modeling of extreme values.* New York: Springer Verlag.
Coles, S.G., and J.A. Tawn. 1991. Modeling extreme multivariate events. *Journal of the Royal Statistical Society, Series B* 53:377–392.
Coles, S., J. Heffernan, and J. Tawn. 1999. Dependence measures for extreme value analysis. *Extremes* 2(4):339–365.
Cooley, D., J. Cisewski, R.J. Erhardt, S. Jeon, E. Mannshardt, B.O. Omolo, and Y. Sun. 2012. A survey of spatial extremes: Measuring spatial dependence and modeling spatial effects. *Revstat* 10:135–165.
Dole, R., M. Hoerling, J. Perlwitz, J. Eischeid, P. Pegion, T. Zhang, X.-W. Quan, T. Xu, and D. Murray. 2011. Was there a basis for anticipating the 2010 Russian heat wave? *Geophysical Research Letters* 38:L06702. doi:10.1029/2010GL046582.
Gumbel, E.J., and C.K. Mustafi. 1967. Some analytical properties of bivariate exponential distributions. *Journal of the American Statistical Association* 62:569–588.

Hansen, J., M. Sato, and R. Ruedy. 2012. Perception of climate change. *Proceedings of the National Academy of Sciences* 109(37):E2415–E2423. Available: http://www.pnas.org/cgi/doi/10.1073/pnas.1205276109 (accessed October 4, 2012).
Herweijer, C., and R. Seager. 2008. The global footprint of persistent extra-tropical drought in the instrumental era. *International Journal of Climatology* 28(13):1,761–1,774. DOI:10.1002/joc.1590. Available: http://www.ldeo.columbia.edu/res/div/ocp/pub/herweijer/Herweijer_Seager_IJC.pdf (accessed October 4, 2012).
Kharin, V.V., and F.W. Zwiers. 2000. Changes in the extremes in an ensemble of transient climate simulations with a coupled atmosphere-ocean GCM. *Journal of Climate* 13:3,760–3,788.
Lau, W.K.M., and K.-M. Kim. 2012. The 2010 Pakistan flood and Russian heat wave: Teleconnection of hydrometeorological extremes. *Journal of Hydrometeorology* 13:392–403.
Min, S.-K., X. Zhang, F.W. Zwiers, and G.C. Hegerl. 2011. Human contribution to more-intense precipitation extremes. *Nature* 470:378–381.
Pall, P., T. Aina, D.A. Stone, P.A. Stott, T. Nozawa, A.G.J. Hilberts, D. Lohmann, and M.R. Allen. 2011. Anthropogenic greenhouse gas contribution to flood risk in England and Wales in autumn 2000. *Nature* 470:302–306.
Rahmstorf, S., and D. Coumou. 2011. Increase of extreme events in a warming world. *Proceedings of the National Academy of Sciences* 108(44):17,905–17,909.
Ramos, A., and A. Ledford. 2009. A new class of models for bivariate joint tails. *Journal of the Royal Statistical Society, Series B* 71:219–241.
Ramos, A., and A. Ledford. 2011. An alternative point process framework for modeling multivariate extreme values. *Communications in Statistics—Theory and Methods* 40(12):2,205–2,224.
Stott, P.A., D.A. Stone, and M.R. Allen. 2004. Human contribution to the European heatwave of 2003. *Nature* 432:610–614.
Wehner, M.F., R.L. Smith, G. Bala, and P. Duffy. 2010. The effect of horizontal resolution on simulation of very extreme U.S. precipitation events in a global atmospheric model. *Climate Dynamics* 34:243–247.
Zwiers, F.W., X. Zhang, and Y. Feng, Y. 2011. Anthropogenic influence on long return period daily temperature extremes at regional scales. *Journal of Climate* 24:881–892.

Appendix E

Foundations for Monitoring Climate–Security Connections

Conclusion 6.1 sets out the five types of key phenomena for which monitoring to anticipate national security risks related to climate events is needed:

1. Climate and other biophysical environment phenomena;
2. The exposures of human populations and the systems that provide food, water, health, and other essentials for life and well-being;
3. The susceptibilities of people, assets, and resources to harm from climate events;
4. The ability to cope with, respond to, and recover from shocks; and
5. The potential for outcomes of inadequate coping, response, and recovery to rise to the level of concern for U.S. national security.

There is substantial variation in the degree to which the expert communities that study and analyze these phenomena have achieved consensus about a set of key variables that are essential for monitoring. In many cases the foundation for identifying, collecting, and analyzing basic data is still under construction. The sections in this appendix reflect the committee's judgment about some of the major data and monitoring needs and capacities related to the five types of key phenomena above. In addition, as discussed in Chapter 6, there are significant challenges associated with developing and applying the analytic techniques needed to turn ever-increasing amounts of data into useful information.

CLIMATE AND OTHER BIOPHYSICAL ENVIRONMENT PHENOMENA

The effects of human activity on the natural environment coupled with natural changes in the physical climate system make long-term monitoring of the global climate system crucial both to understanding the variability and changes in the Earth system and for providing inputs to model-based prediction schemes. As the human population continues to increase, so too will global demands on agricultural, water, and energy resources. Analyzing regional climate impacts and assessing human vulnerabilities will require high-frequency and spatially dense observations as well as information on the change and rate of change of the global climate system. Regional and national networks must be developed, particularly in regions currently experiencing an increased demand on natural resources. If sufficient observations are not collected, the ramifications will be serious, including less accurate weather forecasts and an inability to monitor natural hazards.

As discussed in Chapter 6, in 1998 the Intergovernmental Panel on Climate Change (IPCC) and the United Nations Framework Convention on Climate Change (UNFCCC) established specific requirements for systematic climate observations and a sustained observing system. Those requirements included supporting research to understand more fully the causes of climate change, to predict future global climate change, and to characterize extreme events important to impact assessments, adaptation, risks, and vulnerability. In 2003 the Second Adequacy Report on the Global Observing System (Global Climate Observing System, 2003) concluded that while improvements had been made in the global observing system, deficiencies remained in the global coverage and quality of ocean, atmosphere, and terrestrial measurements. The report concluded that satellite observations over all domains were essential to the global observing system and that they must continue uninterrupted. Despite such urging, however, U.S. satellite observation capabilities are expected to decline by 25 percent over the next 8 to 10 years, according to a recent National Research Council (NRC) report (National Research Council, 2012) on NASA's implementation of the Decadal Survey. (See further discussion below.)

The Global Climate Observing System, sponsored by the World Meteorological Organization, the United Nations (UN) Educational, Scientific and Cultural Organization, the UN Environmental Program, and the International Council for Science, is charged with advising the community on global climate observations and overseeing implementation based on UNFCCC standards. In 2010 the organization developed a list of 50 essential climate variables (ECVs) that are possible to implement globally and whose observation could yield significant progress toward meeting the UNFCCC requirements (Global Climate Observing System, 2010). These ECVs are concerned with atmospheric conditions over land, sea, and ice and include such variables as temperature, wind speed and direction, pressure, cloud properties, and carbon dioxide levels. Ocean ECVs include sea surface and subsurface temperature and salinity, ocean color, sea ice, sea level, oxygen, and nutrients. Most terrestrial observations focus on ground water, water use, snow cover, land cover, and soil moisture.

The ECVs represent a consensus on a broad and comprehensive set of parameters to document Earth's climate system. Designed primarily to characterize key aspects of the Earth system, they are aimed more at informing scientific analysis more than policy decisions. There have, however, been several attempts to develop indices based on the ECVs that would yield an integrated measure of climate impacts that could be more useful for policy analysis. One such index is the U.S. Climate Extremes Index (CEI). It was developed as a "monitoring and communications tool to help U.S. citizens and policymakers identify possible trends or long-term variations in a variety of climate extremes indicators" (Gleason et al., 2008). The CEI is composed of five parameters: monthly minimum and maximum temperatures, daily precipitation, days with and without precipitation, and the Palmer Drought Severity Index (PDSI) (Palmer, 1965). Temperature is important for monitoring a variety of phenomena, including heat waves, cold waves (including freeze events such as late spring freezes), and even unusually warm or cold months and seasons that can have effects on a variety of sectors. Precipitation is the basic building block for monitoring precipitation deficits and excess, including drought and heavy rain events. Air pressure is important for monitoring storms, heat waves, and other events, while water vapor is important for monitoring the potential for heavy rain events and drought. The PDSI is a dryness indicator based on a combination of recent temperature and soil moisture observations.

The CEI was tailored for the continental United States. Currently there is an effort to develop a CEI that would provide a more globally integrated picture of the current state of climate extremes. However, the development of such a global CEI has been hampered by a lack of data availability in many regions and for most of the ECVs. A white paper prepared for the 2011 World Climate Research Programme Open Science Conference (Trenberth et al., 2012) concluded that although the data for ocean and atmospheric ECVs have adequate global coverage over a long enough time and sufficient quality, data for terrestrial measurements are seriously deficient. It also concluded that although existing in-situ data cover most of the high-priority regions, spatial and temporal coverage could be improved. It called for more integration of satellite data as well as for new observations that would provide information about such areas as climate change mitigation and adaptation efforts. Data at regional and local scales on such factors as soil moisture, stream flow, and sea surface temperature that are related to climate events, are extremely inadequate. Soil moisture, for example, is important for monitoring drought as well as for estimating the potential for flooding. Streamflow is a good large-area integrator of short- and longer-term moisture conditions for a basin, a region, or even a continent. Sea surface temperature is especially important for tracking El Niño–Southern Oscillation events.

The white paper found that such data, as well as detailed hourly information on variables such as precipitation intensity, distribution, frequency, and amount of precipitation, are necessary for predicting extreme events on regional scales. Because there are not yet adequate data in these areas, it may be necessary to develop global indicators in an incremental fashion—for instance, by adding one or two parameters at a time or by beginning with specific regions that allow early development, such as Europe or Australia.

Numerous analyses have documented the linkages between global climate change and environmental sustainability (e.g., National Research Council, 2010a, 2010b, 2010c). One example is the way in which changes in ocean and atmospheric circulation force corresponding changes in ocean temperatures, which reach the ice shelves, resulting in land ice loss, sea level rise, and, ultimately, coastal erosion. Likewise, changes in precipitation patterns can affect the snowpack and surface hydrology, thereby affecting agricultural productivity. And increased carbon dioxide emissions absorbed by the ocean lead to increased ocean acidity, which destroys the marine organisms that provide food sources for other marine life and thus negatively affects fisheries. To understand these relationships it is necessary to measure a broad and diverse set of the variables that connect global climate change with human life-supporting systems such as those providing for food, energy, water, and health. Global-scale indicators and metrics based on a broad spectrum of observations can provide some advance warning of the impacts of global climate change (National Research Council, 2010b). Sea level rise is one such indicator, for example, because it is a function of oceanic, land ice, and hydrologic processes. However, to obtain better estimates and projections of rising sea levels it will be necessary to make better sustained observations of such variables as sea state, atmospheric wind speed and direction, sea-ice extent, the mass balance of mountain glaciers and ice sheets, and river discharge.

The United States and other countries do not currently have national strategies for sustaining long-term environmental observations. Joint efforts in support of the UNFCCC and IPCC are encouraging international collaboration, and many nations have recognized the need for a fully implemented observing system. However, funding is a major obstacle. Major gaps in satellite and in-situ observations are developing because of the loss of several key satellites, such as Cryosat, the Orbiting Carbon Observatory, and Glory, and because of the reduced maintenance on the Tropical Ocean Atmospheric Array of ocean buoys and on land-based carbon towers. Delays or cancellation of other satellites will further increase the data gap. Lack of data will severely impair the ability of climate scientists to understand and forecast changes from interactions and feedbacks in the Earth system and will limit the information available to users and decision makers. Serious consequences will include less accuracy in forecasting and in the projecting of natural hazards and extreme events; as these events affect a growing population that is already placing significant stress on Earth's systems and is expanding into areas exposed to likely climate hazards, the consequences will only grow with time.

The national and international efforts noted above have focused mainly on understanding and anticipating climate changes, climate events, and some of their direct consequences. The field has made progress in setting priorities, but its emphasis has not been on security issues. Such an emphasis would require types of monitoring not mentioned above. A considerable number of the types of monitoring relevant to security issues were identified in a previous report from the NRC, *Monitoring Climate Change Impacts* (National Research Council, 2010b). We note that many of the environmental variables that are important to human life-supporting systems cannot be measured readily or with sufficient resolution and accuracy from satellites and instead require place-based measurement and automated interrogation and assimilation. Examples of such variables include the quality of agricultural land, the condition of aquifers, and the ranges of key pests and pathogens.

In addition, to our knowledge there has as yet been no serious priority setting among environmental measurements, looking at the issue through the lens of security analysis. Thus the priorities noted here and in the previous NRC effort (National Research Council, 2010b) provide only a starting point toward prioritizing climate and environmental monitoring needs for security analysis purposes. Efforts to develop priorities should aim at identifying a small number of composite indices designed for the specific purposes of analysis or early warning. Progress will require additional work, which should be conducted through collaborations involving climate scientists, environmental scientists, social scientists, and security analysts. An effective monitoring system is feasible in principle, but it will require an authoritative setting of priorities, the integration of climate and social and political stress indicators, and the development of reliable protocols for international collaboration.

EXPOSURE TO POTENTIALLY DISRUPTIVE CLIMATE EVENTS

Exposure refers to the presence of people, livelihoods, environmental services and resources, infrastructure, or economic, social, or cultural assets in places that could be adversely affected (Intergovernmental Panel on Climate Change, 2012:3). For security analysis it is important to recognize that different climate hazards matter for different places because each population and location of interest has a particular pattern of exposures and susceptibilities to harm as well as of abilities to cope and respond. Therefore, monitoring of exposures needs to include both changes in the frequency and intensity of potentially disruptive climate events and changes in the societal conditions that affect which people and things are exposed to harm from specific events and the different exposures of different segments of societies. This section focuses on the monitoring of people and things. It is, of course, equally important to monitor the hazards. Although some hazards, such as large-scale floods, are easy to monitor, others (e.g., droughts) are not presently well monitored. The following are, in our judgment, some of the monitoring requirements that are most important for estimating exposures to potentially disruptive climate events.

Population Estimates on Spatially Disaggregated Scales

Understanding how hazards might affect human societies requires a basic understanding of where people are located with respect to the hazards, and estimates of exposed populations form a fundamental building block of any vulnerability assessment. The most basic indicator is the number of people residing in a given area, sometimes called the *headcount*. In some cases one wants to know more demographic detail than simply the headcount, for example, the age structure, the number of orphans, the percentage of female-headed households, and so on. Such information is often available in national censuses.

In some cases the most important variable to estimate is not the number of residents in an area, but the total number of people located in the area at a given time, regardless of their residence. This total number of people in an area, including both residents and nonresidents, is sometimes referred to as the *ambient population*. For example, many of the people exposed to the Asian tsunami in 2004 were tourists, who were not counted in the local censuses. In other cases a region may have large numbers of migratory workers whose exposure is relevant to understanding risk but who are not counted in the censuses. An example is China prior to the 2008 financial crisis, when a large number of people located in vulnerable areas along the coast were migrant workers from the interior.

In many climatically fragile areas, people respond to shifting climatic patterns by taking part in seasonal migration. Some massive population movements (e.g., the Hajj) also take place at times of major holidays. Such dynamics can not only affect estimates of exposure, but also influence the emergence of a hazard (e.g., infectious disease). For some hazards it may be helpful to estimate population distributions on an even finer time-scale (e.g., hourly); in some large urban areas the number of people present varies by orders of magnitude over the course of a day.

There are a handful of databases that have been used to generate estimates of these phenomena. Of these the Gridded Population of the World (GPW), produced by Columbia University, and LandScan, produced by Oak Ridge National Laboratory, are the most widely used. Regional collections are also available. (For a recent review see Tatem et al., 2012.) GPW integrates national census data using the highest spatial resolution available; LandScan estimates ambient population using census data along with spatial correlates of population location.

Such information is useful across the entire life cycle of hazard assessment, preparedness, and response. Well before an event of concern, such information is relevant for risk assessment and planning for reducing susceptibility among exposed populations. As an expected event approaches, the information is relevant for preparedness and crisis response planning (evacuation, resource mobilization, etc.). And in post-crisis periods, it is relevant for damage assessment and reconstruction planning; at this stage, census-derived data are no longer relevant, and estimates must be made afresh using ground surveys and satellite imagery.

The main weaknesses of the available data are the temporal lags (decennial censuses are inadequate for many purposes) and the lack of spatial resolution.

Location and Characteristics of Critical Infrastructure

Some risks are more a function of the exposure of critical infrastructure than of the exposure of populations. In such cases, it is important to monitor the locations and key characteristics of such infrastructure as power plants, roads, railroads, ports, telecommunications centers, hospitals, prisons, government buildings, and key manufacturing facilities. It is important that, in addition to monitoring, there should be an ongoing assessment of the adequacy of this infrastructure and of whether plans are in place for backup procedures if the adequacy standards are not yet met.

Publicly available global databases on such infrastructure are not considered adequate. Data on roads, for example, are available but are outdated and of very low accuracy, while other infrastructure information is not available except in commercial proprietary databases (mainly from the insurance industry and large engineering firms) and from classified military databases. There is an effort under way to compile a global exposure database for use in earthquake modeling[1] that will represent the most comprehensive public database of this type when complete. However, because it is oriented toward modeling economic loss rather than modeling security threats, it cannot be expected to meet all needs for security assessments.

Information on infrastructure exposures would be useful at all stages of the life cycle of potentially disruptive events: early on for risk assessment and planning, and later on for damage assessment and response prioritization.

Land Use and Land Cover

Patterns of land use affect vulnerability to hazards and also constitute an integral aspect of exposure. Whether a given area is devoted to commercial agriculture, subsistence agriculture, grazing, dense urban settlement, or wilderness will make a difference on how a given hazard affects security dynamics. For example, a prolonged drought in an area with significant subsistence agriculture and grazing will likely result in population movements, whereas such a response would be less likely in an area with other land uses.

There are standard classification schemes for characterizing land use and land cover, and these form the basis for generating indicators. Indicators that characterize different types of agricultural activity and different levels of urbanization would be particularly useful.

A number of global databases measure land use and land cover, with satellite imagery providing the most important input, along with significant validation from ground-based observations. There are a number of sources of global data on land cover change, including NASA's Land-Cover/Land-Use Change Program,[2] the Global Land Cover Facility at the University of Maryland,[3] and the U.S. Geological Survey.[4] Other sources provide land use data for specific counties or regions. For example, the Department of Global Ecology of the Carnegie Institution for Science, based at Stanford University, compiles high-resolution data on forest cover and related variables for tropical regions.[5]

These databases are considered accurate enough for use in global modeling and other broad exercises, but their accuracy and precision within specific regions is low. Their precision is low enough, for example, that they

---

[1] GED4GEM. See http://bit.ly/KCzpyj (accessed November 15, 2012).
[2] See http://lcluc.umd.edu/data_information.php (accessed November 15, 2012).
[3] See http://esip.umiacs.umd.edu/index.shtml (accessed November 15, 2012).
[4] See http://landcover.usgs.gov/index.php (accessed November 15, 2012).
[5] See http://claslite.ciw.edu; http://cao.stanford.edu (accessed November 15, 2012).

tend not to discriminate between subsistence and commercial agriculture or between different levels of urbanization. They also tend to have difficulty in characterizing pasture, a critical land use category for understanding climate impacts. Moreover, the databases tend to disagree widely in delineating urban extents.

Data on land use and land cover are most useful when used for risk assessment and planning in the early stages of the life cycle of potentially disruptive events.

SUSCEPTIBILITY TO HARM

Susceptibility is the likelihood of immediate harm to a population, community, society, or system as the result of exposure to a climate event. Thus, susceptibility is an indicator of the extent that an event would create disruptive change in the short term in that population, community, society, or system. For security analysis, it is important to identify factors that influence the general susceptibility of populations to all types of climatic shocks and stresses as well as the susceptibility to specific types of climate-induced events, such as a flood or a pandemic. This section discusses monitoring needs for measures of general susceptibility before turning to a few key areas of specific susceptibility: food, water, and health.

Monitoring of General Susceptibility

A number of factors influence general susceptibility, including various economic, demographic, social, and environmental conditions in a region; the form and quality of the infrastructure and the built environment (Intergovernmental Panel on Climate Change, 2007, 2012); and the presence or a recent history of violent conflict (Barnett, 2006; Barnett and Adger, 2007; Brklacich et al., 2010). In most cases it is not only the values of the variables that are relevant but also the direction and rate of change in those variables. For example, while the level of poverty in a region is an indicator of general human well-being and therefore of susceptibility to harm, there is additional value in knowing whether the level of poverty is increasing. If it is—even if poverty in the population does not seem particularly high to start with—this would suggest that well-being is deteriorating and that the potential for harm from climatic shocks may be increasing because financial and other assets that previously would have allowed households to cope with exposure to climate shocks are being depleted and because the construction of protective infrastructure is lagging far behind and further exposing populations to risk (Parry et al., 2009).

In addition to determining which types of susceptibility factors should be monitored, it is also critically important to consider the appropriate scale or level of analysis for assessing and monitoring the susceptibility of specific populations. A susceptibility assessment may apply to an entire population or to a defined subset of a population (such as a particular ethnic group) within a politically bounded region, such as a city, state, or country, or within various types of functional regions, such as urban neighborhoods, border zones, agricultural areas, hydrologic basins, and so forth. As a general rule, susceptibility monitoring should emphasize locations that are currently or potentially of security concern, but it should also pay attention to regions where major humanitarian or other types of crises may arise or where large migration inflows or outflows may be likely.

It is also important to recognize that many of the factors that influence susceptibility to future climatic shocks and stress are already in flux as the result of ongoing environmental and climatic changes (Paavola, 2008) and of non-climatic processes, including globalization and urbanization (O'Brien and Leichenko, 2007; Leichenko and O'Brien, 2008). For example, rapid rates of deforestation on the hill slopes surrounding a city, which are often associated with population immigration, will exacerbate the susceptibility of a region to extreme precipitation events. Areas where susceptibility indictors are rapidly changing—especially when these changes suggest that susceptibility is increasing—also merit special attention because these are regions where social and political turmoil may be more likely, particularly if the regions have a recent history of violent conflict.

Measurements of economic conditions provide an indication of the financial and material well-being within a population or region and of the capacity to withstand climate-related shocks and stresses. Regions or population subgroups with low or deteriorating levels of per capita income and financial assets, high levels of poverty, high levels of unemployment, or high levels of income inequality can be expected to be more susceptible to harm from climatic risks and hazards. Economic diversity also has an important influence on susceptibility. Regions or populations that depend on a single agricultural commodity for their livelihoods and where alternative livelihood options are limited also tend to be more susceptible to harm from all types of climatic shocks and stresses, particularly those affecting their main source of livelihood. Potential indicators of economic well-being and economic diversity include per capita income, poverty rates, levels of inequality, unemployment rates (overall and by age cohort), percentage of the labor force in agriculture, and share of agricultural production by crop.

Many of these basic economic variables are regularly catalogued by international organizations such as the World Bank, which maintains databases on variables at the level of the nation-state for most countries.[6] National-level information on agricultural production and crop yields is

---

[6]See http://databank.worldbank.org/data/home.aspx (accessed November 15, 2012).

available from the International Food Policy Research Institute[7]; other data sources are discussed later. Many national governments also collect economic data at both the national and subnational levels, but for less-developed countries the availability and reliability of these data are limited, and many countries do not release their routinely collected economic data. Subnational data are also available in datasets of the Center for International Earth Science Information Network at Columbia University; these datasets contain subnational estimates of indicators of economic well-being, including poverty, inequality, unmet basic needs, and food security, for a large number of countries.[8] One important limitation of using existing data from many publicly available secondary sources is that the underlying source data, such as national population censuses, are updated relatively infrequently. These datasets provide reasonable estimates of baseline conditions, but they would need to be supplemented with primary data collected for specific regions of interest in order to gauge current conditions or to assess changes over shorter periods of time.

Demographic and social conditions are another group of factors that influence susceptibility to harm from climatic shocks and stresses. Regions with high rates of population growth, particularly from immigration, can be expected to be more susceptible to many climate risks because, as discussed in Chapter 5, new immigrants to a region are often poorer than long-time residents, are likely to lack knowledge of local environmental conditions, and tend to live in hazard-prone areas such as flood plains. Highly urbanized areas concentrate people and may therefore increase the disruption that would result from events experienced there; on the other hand, it may be easier in highly urbanized areas for response efforts to reach affected people. Regions with high shares of elderly residents or of young children also tend to be more susceptible to multiple hazards because these groups rely on others for their safety and well-being. Regions with low levels of education and literacy—both of which are indicators of human capital—would also be expected to be more susceptible to harm from a variety of climate events (Cavallo and Noy, 2010). In regions with high levels of gender inequality, females could be expected to be more susceptible to harm from climate events than males because of a relative lack of assets and a lack of access to resources that may be available to men. The health status of populations also influences susceptibility to harm from such hazards as food insecurity, poor water quality, and exposure to infectious disease agents.

Potential demographic and social variables to monitor include population growth rate, rate of immigration, share of the population that is over

---

[7] See http://www.ifpri.org/dataset/agro-maps-mapping-agricultural-production-systems (accessed November 15, 2012).
[8] See http://sedac.ciesin.columbia.edu/theme/poverty (accessed November 15, 2012).

age 65, share of the population that is under age 12, education levels, literacy rates, and female literacy rates. As with the economic variables listed above, many of these demographic indicators are regularly cataloged at the national level by international organizations and at the subnational level by national governments. As discussed in the section on exposure, detailed population estimates at the subnational level are also available via satellite-based datasets. However, subnational coverage for indicators of dependency, literacy, and gender inequality are limited for developing countries and may require new data collection efforts, such as household surveys, in regions of security concern.

Another factor that influences susceptibility to extreme climate events is the built environment and physical infrastructure, particularly for urban populations in developing countries. Many of the variables that are relevant to exposure, such as the quality of housing and building stock and the presence and pattern of informal housing settlements, also influence the amount and type of physical damage done and the loss of life that occurs following exposure to climatic events. The characteristics of a city or urban region's physical infrastructure, particularly the maintenance of piped water supplies, transportation networks, and electricity grids, all influence the susceptibility of the region's population to harm from extreme climate events affecting that infrastructure. Poorly maintained water supply systems are more likely to be contaminated during flood events; improperly maintained bridges are more likely to collapse or experience significant damage when exposed to extreme weather; and inadequate or poorly maintained electrical and communication grids may be down for longer-than-normal periods following exposure to extreme events. The most important variables to monitor and the best data sources for characterizing the built environment and physical infrastructure were discussed in the section on exposure.

A region's natural capital, including its natural physical assets and ecosystem services, influences the quality of life and livelihood options in a region. As discussed below in relation to food security and water security, regions with degraded or deteriorating environmental conditions tend to be more susceptible to harm or disruption from various climatic events. In addition to the environmental factors associated with water and food security, such as soil degradation and the quality and reliability of freshwater supply, other indicators of the conditions of natural assets include the rate of deforestation in a region and the rate of wetland or mangrove loss. Sources of data for the latter were discussed above.

For all types of extreme climate events the capacity of a region's governing institutions and the presence of social capital influence the likelihood that a population will be harmed. Some potential areas to monitor include the density of social networks, the level of public spending per capita, the transparency of governing institutions, and the degree of corruption in governing institutions. Data sources and limitations are discussed in the section on coping.

The presence or recent history of conflict is another important indicator of susceptibility to harm from climatic shocks and stress. Conflict damages infrastructure and life-supporting systems and undermines the capacity of institutions to prepare for and respond to climatic hazard events (Barnett, 2006; Barnett and Adger, 2007; Brklacich et al., 2010). Populations living in regions where conflict is present are highly susceptible to harm from climate risks and hazards. Possible variables to monitor and data sources are described in the section below on the stability and fragility of social and political systems.

Monitoring Water Security

The section on exposure earlier in this chapter addressed the monitoring of some of the climate-related variables that affect water supply along with other aspects of physical geography. Measurements of these variables will provide some of the baseline conditions needed for identifying and tracking changes. Generally speaking, climate change and climate events will produce conditions of both too little water (droughts) and too much (floods), and these will interact with the physical baseline conditions. Assessing susceptibility requires the additional consideration of the many ways in which human activity will affect the baseline. Many of those factors, such as increasing population and urbanization and changing patterns of agriculture and food consumption, would be part of monitoring general patterns of susceptibility and are relevant to water security if the measurements match the locations where water security is an issue. As discussed in several chapters, this combination of physical and socioeconomic trends means that, over the next decade and beyond, many regions are likely to face significant water challenges, including shortages, insufficient quality, and competition over water for different purposes. In discussions about monitoring water security it may be helpful to make a distinction between use-related variables (both demand and use) and water-management–related variables.

Use-related variables. Water use can be measured in a number of ways, and care must be taken in comparing different metrics. The quantity of water withdrawn or consumed from surface water and groundwater for human purposes is the most frequently used measurement of water use. Some withdrawn water may be returned to the surface or to groundwater after use; water consumption refers to the water that, for one reason or another, is not returned to the system. Other considerations of water use are its availability when needed and its quality with respect to different purposes.

A measure that captures both water supply and demand and that underpins several concepts is water supply per capita per year, which divides the total average renewable water supply in an area by its population. *Water stress* is defined as the situation when a country or region reaches a point at which annual water supply is less than 1,700 cubic meters per person per year (Office of the Director of National Intelligence, 2012; for reference, in the United States the total amount of water used per capita is 2,500 cubic meters per year). *Water scarcity* occurs when a country's or region's annual water supply is less than 1,000 cubic meters per person per year.

It is not yet possible to obtain comprehensive and consistent data on water availability and use for most parts of the world (Gleick, 2011). The UN Food and Agriculture Organization maintains the Aquastat database, which assembles information about "water resources, water uses, and agricultural water management from a variety of sources, with an emphasis on Africa, Asia, Latin America, and the Caribbean."[9]

The degree of access to safe drinking water and sanitation provides one basic measure of water development and conditions as well as of societal capacity. Because one of the UN's Millennium Development Goals is major improvements in access to drinking water and sanitation, there are regular reports about progress toward that goal (United Nations, 2012). There are also a number of sources of data on access to safe drinking water and sanitation, such as the Demographic and Health Surveys, a project of the U.S. Agency for International Development (USAID).[10]

The 2010 NRC report *Monitoring Climate Change Impacts* proposed six hydrology variables to assess climate-change impacts on water supply as well as measurement approaches and rationales for their links to environmental sustainability. The six variables are the volume (or mass) of water stored in parts of the terrestrial system; seasonal snow cover and snow water equivalent, and their seasonal progression; fluxes of water through the land–water system; water quality; lake, river, and reservoir ice cover; and the contribution to sea level rise from land water storage (National Research Council, 2010b). Complementing these variables with others that capture how water supply and demand are managed would provide the capacity to monitor water security.

Management-related variables. By "management-related variables" we mean both those variables that reflect the management of "internal" water supplies—that is, those that fall within the jurisdiction and control of national governments—and also variables related to transboundary water management issues. Both internal and transboundary supplies, and the populations that rely on them, are susceptible to harm from climate change and events. The monitoring of national and transnational water manage-

---

[9] See http://www.fao.org/nr/water/aquastat/main/index.stm (accessed November 15, 2012).
[10] See http://www.measuredhs.com/Who-We-Are/About-Us.cfm (accessed November 15, 2012).

ment will rely in part on qualitative analysis and expert judgment, and it could complement the more general monitoring of government capacity and patterns of conflict and cooperation.

In monitoring internal water management, one would want to have information about the components of the water management system as well as about its capacity. The former category would include how decisions about pricing and allocation are made, the current conditions of the water infrastructure (e.g., dams, levies, canals, and water treatment facilities) and of projected developments, and how the water infrastructure is used to manage water flow. The capacity of the system could be gauged by such factors as current and projected investments in technology or management improvements, especially those intended to address scarcity, and assessments of the quality and legitimacy of pricing and allocation. It could be also useful to know how politicized water issues are in a region, which could be assessed by monitoring popular media using either events data techniques or the newer technologies to capture social media.

The components of an assessment of transboundary water management would include the number and type of agreements about the use of water as well as some of their specific components, such as allocation mechanisms and procedures for dispute resolution. In assessing capacity, it could be helpful to track patterns of cooperation or conflict over water to identify anomalies or emerging trends. As with the political dimensions of internal water management, it could be helpful to understand how elites frame water issues and to determine the nature and level of politicization.

There are several sources of information that can be used in the monitoring of some of the management issues. One of the best known is the Water Conflict Chronology of the Pacific Institute, with data on cases from 3000 BCE to 2010.[11] Another is the International Water Events Database, maintained by the Institute for Water and Watersheds at Oregon State University, which captures various events reported in the media between 1950 and 2008 and codes them along a conflict–cooperation scale from −7 to +7.[12] Both databases capture internal as well as interstate events. As discussed in Chapter 6, the investments being made in improving the capacity to monitor media sources, such as including more types of sources in more languages, could aid in identifying emerging patterns of conflict and cooperation earlier and in more specific areas.

---

[11] Current, sometimes overlapping categories of types of conflicts now include measures that may be relevant to questions of future changes, such as increases in terrorist activity (Water Conflict Chronology: http://www.worldwater.org/conflict.html [accessed June 23, 2012]).

[12] For more information, see http://www.transboundarywaters.orst.edu/database/interwater eventdata.html (accessed November 15, 2012).

> **BOX E-1**
> **Environmental Indications and Warnings**
>
> Freshwater is an essential resource for civilization that is tightly coupled to climate and weather. We live in a world today that is significantly more water stressed than at the beginning of the century, and we can expect it to get worse. Demand for freshwater has increased approximately eight-fold in the past century because of population growth and economic development. At the same time, climate changes are affecting the annual renewable supply and distribution of freshwater by altering temperature and precipitation regimes.
> These changes have been recognized as a security concern by the U.S. National Intelligence Council, which concluded that "during the next 10 years, water problems will contribute to instability in states important to US national security interests" (Office of the Director of National Intelligence, 2012:iii). In order to stay abreast of potential threats to U.S. national security posed by emerging freshwater security issues, an early warning capability called Environmental Indications and Warnings (EIW) has been developed to monitor a wide range of climate and water security indicators in order to identify troublesome patterns and vulnerable regions.
> Each month EIW issues global reports to analysts that provide localized information about current and historical anomalous conditions, with projections nine months into the future, by using the National Oceanic and Atmospheric Administration seasonal climate forecasts to forewarn of possible events:
>
> 1. *Synoptic global "hot spot" maps* that provide the ability to spot emerging issues at subnational scales that deserve a more detailed analysis (see Figure E-1, right);
> 2. *Water security case studies* that provide regional historical context with which to assess how current freshwater conditions compare with past events and to determine which pose a societal threat;
> 3. *Detailed sectoral assessments* of how freshwater stress is affecting pastoral and agricultural lands, electrical power generation, and urban water systems; and

Box E-1 describes a major U.S. government–funded project to monitor a wide range of climate and water security indicators in order to identify troublesome patterns and vulnerable regions.

Monitoring Food Security

In the domain of food security, the important conditions to monitor include growing conditions for the food crops that are important to a region (whether grown locally or imported), production levels for these crops,

APPENDIX E

4. *Statistical analysis* to assess trends in the prevalence of extreme events and also the relationships between freshwater stress and specific national security–relevant outcomes.

EIW systematically monitors acute and persistent freshwater anomalies, integrates measures of environmental stress with political and socioeconomic factors, calculates a suite of indicators that reflect place-based sensitivity to stress, and provides analysts with easy-to-interpret products. While EIW draws extensively on open scientific research, it processes this information in a manner that is driven by and uniquely suited to the needs of national security analysts.

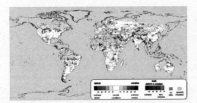 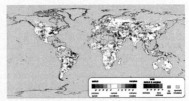

FIGURE E-1 Hot spot maps: Composite freshwater surpluses and deficits. These maps present composite pictures of freshwater surpluses and deficits for the three years from July 2009 through June 2012 (left) and the three months from April through June 2012 (right). The three-year period is useful for highlighting regions with persistent deficits and surpluses, and the three-month period is useful for highlighting short-term anomalies. The colors yellow through dark red represent increasingly severe deficits, while cyan through dark blue represent increasingly severe surpluses. Shades of purple are used to depict regions that exhibit aspects of both surplus and deficit at the same time. Note, for example, that the region along the border between the United States and Mexico is experiencing both short-term and persistent drought.

SOURCE: Material provided by the U.S. government.

grain storage levels, susceptibility to large price increases in food or energy prices, global prices of major food commodities, local food prices, the incidence of new strains of crop and livestock diseases, and the nutritional status of populations. With regard to the impact of extreme events, one would want to monitor both supply-disrupting natural disasters in net food exporter countries (e.g., the United States, the European Union, Canada, Russia, Australia, etc.), which cause rising food prices and thus increase the likelihood of social unrest abroad, and demand-disrupting natural disasters in net importer countries (e.g., the Middle East and North Africa region), which tend to depress incomes, and thus the ability for people to feed themselves.

It would be useful to monitor several of these conditions to be able to identify potential problem areas months to a year in advance. For example, monitoring grain storage levels would help to foresee times when supply shocks could translate into large spikes in local and global food prices. Specific indicators would include the ratio of global stocks to global annual consumption and the stock/consumption ratios in countries that have limited capacity to import food. Historically, data have been compiled and reported by the U.S. Department of Agriculture,[13] but many analysts question the reliability of these estimates for many countries. Some countries, such as China, have historically had limited interactions with global food commodity markets, so levels of their stocks have been less important for global prices (Dawe, 2009; Wright, 2011); however, that situation has likely changed, at least in the case of China, which is now a major importer of corn and soybeans.

Monitoring of susceptibility to large price increases in food or energy would help to foresee which areas are most likely to experience rapid increases in commodity prices. Important indicators would be the fraction of food consumption derived from imports, the levels of subsidies provided by the government, and the level of governmental financial resources available to sustain those subsidies in the face of rapid price increases. As discussed in Chapter 4, countries in the Middle East and North Africa are particularly dependent on food imports. Many other countries, such as in Western Africa, are dependent both on food and energy imports, and so increases in the price of one could compromise the ability of governments to provide the other. Current efforts to monitor reliance on imports and levels of subsidies and government resources include the World Bank's Development Indicators[14] and the UN's Food and Agriculture Statistical Databases.[15]

A number of other indicators are particularly useful to monitor in order to identify potential problem areas weeks to months in advance. For example, the monitoring of food prices is important for identifying and predicting near-term threats to food security, both in urban areas and also in rural areas where most of the poor are net food buyers. Global commodity prices are a specific class of relevant indicators that provide an easily accessible measure of global food availability. The UN Food and Agriculture Organization (FAO) regularly monitors food prices and produces an

---

[13] See http:// www.usda.gov/oce/commodity/wasde (accessed November 15, 2012).
[14] See http://data.worldbank.org/indicator/TM (accessed November 15, 2012).
[15] See http://faostat.fao.org (accessed November 15, 2012).

aggregate food price index[16]; some researchers have proposed that a food price index above 210 is a strong predictor of unrest throughout the world (Lagi et al., 2011). The International Monetary Fund monitors global prices of all major commodities, including major food grains.[17]

It would also be important to monitor specifically how countries react during episodes of high food prices. Many countries react by imposing export bans on commodities, exacerbating the situation and risking global system shocks. It is more difficult to monitor local food prices than to monitor global prices, but in many areas it would provide a much more reliable indicator of the cost-of-living challenges faced by local populations. The Famine Early Warning Systems Network (FEWS NET; see Box E-2) currently monitors local food prices and other factors relevant to food security throughout many countries in Africa and a number elsewhere (Brown et al., 2012), but its coverage does not extend to a number of areas of interest for U.S. national security.

A related indicator is expectations of future price changes. Interestingly, there is evidence that increased food price volatility actually decreases the likelihood of unrest, presumably because consumers then tend to view high food prices as temporary. It would therefore make sense to monitor the level of food volatility in recent months as well as the tone of media coverage or social network messages about the state of food availability and food prices.

The monitoring of crop and livestock diseases, particularly new diseases to which current crop varieties and livestock breeds are very susceptible, would help identify the potential for rapid and large losses in local—and possibly global—food supplies. A specific indicator could be the reported incidence of known threats by local crop and livestock experts. Some monitoring capabilities do now exist or are in the process of being established. For instance, the emergence of a new wheat stem rust in Uganda in 1999 posed a considerable threat to global wheat supply (Singh et al., 2008), and it spurred a new initiative to monitor global wheat rusts.[18] For livestock, monitoring systems exist in a number of developed countries, but surveillance in many countries is currently limited. Ongoing efforts to improve livestock disease surveillance include those led by USAID[19] and by the Global Early Warning System, a "joint system that builds on the added value of combining and coordinating the alert and disease intelligence mechanisms of OIE [Organization for Animal Health], FAO, and WHO

---

[16]See http://www.fao.org/worldfoodsituation/wfs-home/foodpricesindex/en (accessed November 15, 2012).
[17]See http://www.imf.org/external/np/res/commod/index.aspx (accessed November 15, 2012).
[18]See http://wheatrust.org (accessed November 15, 2012).
[19]See http://www.vetmed.ucdavis.edu/ohi/predict/index.cfm (accessed November 15, 2012).

> **BOX E-2**
> **Famine Early Warning Systems Network**
>
> Created in 1985, the goal of the Famine Early Warning Systems Network (FEWS NET) is to "to lower the incidence of drought- or flood-induced famine by providing to decision makers, timely and accurate information regarding potential food-insecure conditions. With early warning, appropriate decisions regarding interventions can be made."[a] Funded by the U.S. Agency for International Development, the project is a collaboration among national, regional, and international partners, including expert field personnel on the ground.[b] "FEWS NET professionals in Africa, Central America, Haiti, Afghanistan, and the United States monitor and analyze relevant data and information in terms of its impacts on livelihoods and markets to identify potential threats to food security."[c] A range of products provide alerts, monthly status reports, outlooks, and in-depth studies that make up FEWS NET's on-the-ground coverage of 23 countries, mostly in Africa, but also including Haiti, Afghanistan, and countries in Central America. Less extensive coverage is provided for 15 more countries through partner-based monitoring. In all of its locations, FEWS NET has a strong focus on capacity-building and collaboration with local stakeholders in the countries of coverage.
>
> To provide continuous assessments of overall food and water security as well as famine risk, FEWS NET monitors a number of different kinds of data within an overall framework of analysis that divides countries into relatively homogenous livelihood zones representing key features of how people assure their food security (through agricultural production, pastoralism, trade, etc.). FEWS NET relies on a combination of physical and socioeconomic indicators to estimate and predict the degree of, and changes in, the food security conditions of vulnerable countries.

[World Health Organization] for the international community and stakeholders to assist in prediction, prevention, and control of animal disease threats, including zoonoses, through sharing of information, epidemiological analysis and joint risk assessment."[20]

Finally, the nutritional status of populations potentially affected by large decreases in food production or increases in food price is an important indicator of the way such events would affect food security.

Monitoring Health Security

Current and projected climate variability and climate change raise concerns about the population health burdens of a variety of climate-sensitive health outcomes. Those of potential interest to national security include malnutrition, which was mentioned above in discussing food security, and

---

[20]GLEWS: see http://www.glews.net (accessed July 24, 2012).

> FEWS NET was one of the earliest users of satellite imagery to monitor rainfall and crop conditions in the developing world, and it now also looks at longer-term global climate variability[d] to help understand future threats to food security and to inform priorities for climate adaptation activities. FEWS NET's groundbreaking use of food prices and "commodity market networks, market integration, the geographic and economic distribution of food commodities, cross border trade" (Famine Early Warning Systems Network, 2008:1) has led to new ways of seeing the nature of socioeconomic threats to food security in some countries—significantly different patterns of food pricing within countries; strikingly divergent global food price influences in different regions; and local prices that track closely with global prices in some places, but in others they do not. The goal of all this work is to provide humanitarian and developmental decision makers with credible, comparable, and spatially detailed information about current food security, the need for external food assistance, the nature of underlying vulnerabilities, and threats to food and water security along with the opportunities that exist to alleviate these conditions.
>
> [a]U.S. Geological Survey. See http://earlywarning.usgs.gov/fews/dp_overview.php (accessed July 9, 2012).
> [b]FEWS NET is implemented by a team consisting of private firms (Chemonics International, Inc. and Kimetrica, Inc.) and several U.S. government agency partners: the U.S. Geological Survey, the National Aeronautics and Space Administration, the National Oceanographic and Atmospheric Administration, and the U.S. Department of Agriculture. See http://www.fews.net/ml/en/info/Pages/default.aspx?l=en (accessed July 9, 2012).
> [c]FEWS NET. See http://www.fews.net/ml/en/info/Pages/default.aspx?l=en (accessed July 9, 2012).
> [d]See http://chg.geog.ucsb.edu/products/Trend_Analysis/index.html (accessed November 15, 2012).

pandemics such as influenza or yellow fever. Other health outcomes, such as morbidity and mortality due to changes in the frequency, intensity, duration, and spatial extent of extreme events or to changing air and water quality, while potentially significant at local and regional scales, would be unlikely to lead to national security concerns. There may always be exceptions, however. For example, the 2008 cholera epidemic in Zimbabwe affected approximately 92,000 people, with more than 4,000 deaths (Intergovernmental Panel on Climate Change, 2012). The outbreak lasted from August 2008 until June 2009 and ultimately seeded epidemics in other countries. Weather was a crucial factor in the outbreak, with recurring point-source contaminations of drinking water sources amplified by the onset of the rainy reason. The outbreak was primarily centered in urban areas and had a relatively high case fatality rate of 4 to 5 percent (while less than 1 percent is typical). Shortages of medicines, equipment, and staff contributed to the magnitude of the epidemic.

Monitoring is recommended for emerging infectious diseases because

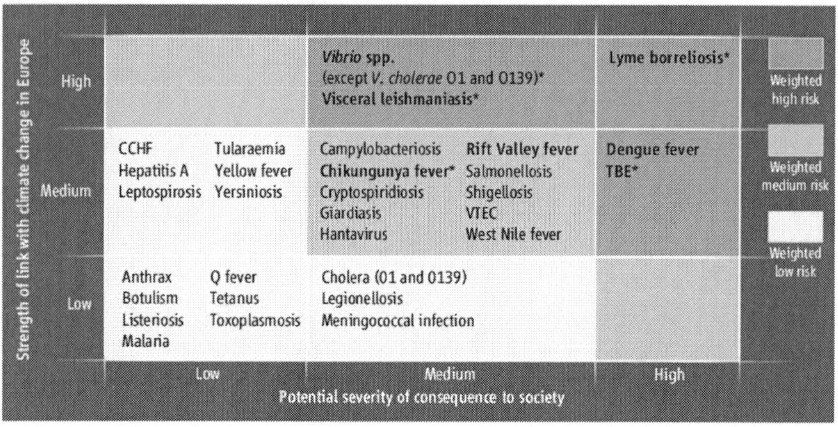

Strength of link with climate change in Europe						
High			Vibrio spp. (except V. cholerae O1 and O139)* Visceral leishmaniasis*		Lyme borreliosis*	Weighted high risk
Medium	CCHF Hepatitis A Leptospirosis	Tularaemia Yellow fever Yersiniosis	Campylobacteriosis Chikungunya fever* Cryptospiridiosis Giardiasis Hantavirus	Rift Valley fever Salmonellosis Shigellosis VTEC West Nile fever	Dengue fever TBE*	Weighted medium risk
Low	Anthrax Botulism Listeriosis Malaria	Q fever Tetanus Toxoplasmosis	Cholera (O1 and O139) Legionellosis Meningococcal infection			Weighted low risk
	Low		Medium		High	
	Potential severity of consequence to society					

Weighted risk analysis of climate change impacts on infectious disease risks in Europe. CCHF, Crimean-Congo hemorrhagic fever. Candidates for suggested changes to disease-specific surveillance are in bold. Asterisks indicate diseases currently notifiable in some EU member states but not legally reportable to ECDC.

FIGURE E-2 Weighted risk analysis of climate change impacts on infectious disease risks in Europe.
SOURCE: From Lindgren et al. (2012). Reprinted with permission from the American Association for the Advancement of Science.

there is a long history of such diseases surprising societies and causing high morbidity and mortality, sometimes accompanied by social disruption (Lindgren et al., 2012). It is not necessary that there be a large number of cases to cause social disruption—as demonstrated, for example, by outbreaks of Ebola. An analysis of climate change–related impacts on emerging infectious diseases of possible concern for Europe is summarized in Figure E-2.

The figure highlights the fact that even well-developed public health infrastructures, such as those in Europe, need modifications to their surveillance and monitoring systems in order to protect their populations from a changing climate. The situation in developing countries is much more challenging. In response to a Presidential Directive in 1996, the U.S. Department of Defense established the Global Emerging Infections Surveillance and Response System, with the mission of monitoring newly emerging and re-emerging infectious diseases among U.S. service members and dependent populations (Clinton, 1996). The U.S. Armed Forces Health Surveillance Center, Division of Global Emerging Infections Surveillance and Response System Operations (AFHSC-GEIS) coordinates a multidisciplinary program to support the International Health Regulations. The goal of the program is to link datasets and information into a predictive surveillance program that generates advisories and alerts on emerging infectious disease outbreaks

(Witt et al., 2011). Datasets and information are derived from eco-climatic remote sensing activities; ecologic niche modeling; and arthropod vector, animal–disease host/reservoir, and human disease surveillance for febrile illnesses. The program includes 39 funded partners working in 92 countries (Russell et al., 2011).

From October 1, 2008, to September 30, 2009, the program identified 76 outbreaks in 53 countries (Johns et al., 2011). The program is characterizing and validating eco-climate anomalies for predictive surveillance for a range of vector-borne and waterborne and zoonotic diseases, including Rift Valley fever, Crimea–Congo hemorrhagic fever, Ebola and other viral hemorrhagic fever outbreaks, malaria, leishmaniasis, and enteric diseases (Fukuda et al., 2011; Money et al., 2011; Witt et al., 2011). Influenza and other respiratory diseases are included in the surveillance activities (Burke et al., 2011). These activities could be enhanced to consider how climate variability and change could alter the risks of outbreaks in geographic regions of interest.

Other organizations monitoring emerging infectious diseases include the U.S. Centers for Disease Control and Prevention; the European Centre for Disease Control; the WHO Global Alert and Response (GAR) system, which includes the Global Outbreak Alert and Response Network; and the Program for Monitoring Emerging Diseases (ProMED) reporting network at the International Society for Infectious Diseases. Diseases covered by GAR include

- Avian influenza
- Influenza
- Crimean–Congo hemorrhagic fever
- Ebola hemorrhagic fever
- Meningococcal disease
- Pandemic influenza (H1N1)
- Rift Valley fever
- Severe Acute Respiratory Syndrome (SARS)
- Smallpox
- Yellow fever

Monitoring these networks can provide insights into changing infectious disease patterns, whether caused by climate change or other factors. In addition, it would be useful to monitor vaccine levels to know whether there will be an adequate supply if a large outbreak should occur. For example, there is an international stockpile of 2 million doses of the yellow fever vaccine,[21] but this would be insufficient if there was a large outbreak.

---
[21] See http://www.who.int/csr/disease/yellowfev/global_partnership/en/ (accessed November 15, 2012).

From 1995 to 2009 dengue fever, a disease carried by the same mosquito that can carry yellow fever, infected 8.88 million people in Latin America and the Caribbean, with 2,870 deaths (Diaz-Quijano et al., 2012). Brazil alone had more than 1 million cases of dengue fever in 2010.[22]

Ideally, the vectors carrying diseases such as yellow fever should be monitored so as to enhance the opportunities for preparedness. However, there is limited consistent and long-term monitoring of most mosquitoes and other vectors, including the yellow fever vector, and, as far as we are aware, there has been no effort to look for range expansions that could develop as a result of climate change (Hayden, 2012).

Monitoring how vectors are distributed around their current range could provide valuable information on emerging risks. For example, Canada has active and passive surveillance of the geographic distribution of the vector that carries Lyme disease, which is now spreading because of warming temperatures and the introduction of ticks to new regions (Ogden et al., 2010). Ogden and colleagues recently developed predictive maps of the geographic spread of the tick *Ixodes scapularis* in order to identify emerging areas of Lyme disease risk (Koffi et al., 2012). Further understanding of the possible impacts of a changing climate and disease transmission dynamics and better long-term datasets could suggest other opportunities for identifying regions at risk of other emerging diseases (National Research Council, 2001). Monitoring *Aedes* mosquitoes, for example, would be helpful for anticipating the spread of yellow fever, dengue, and chikungunya. Monitoring health outcomes during and following extreme weather and climate events, such as flooding and cyclones, could be useful for detecting the beginnings of epidemics of diseases such as cholera.

COPING, RESPONSE, AND RECOVERY—AND THEIR POLITICAL CONSEQUENCES

Key Factors

Given that disruptive natural events are never equal or equitable in their societal impacts and that governments are expected to provide emergency assistance and then to organize, if not lead, disaster recovery, social and political stresses are to be expected in the aftermath of such an event. However, the short-, medium-, and long-term outcomes of those stresses vary greatly, and the different outcomes have significant monitoring implications. As discussed in Chapter 5, the available evidence on the relationships between potentially disruptive climate events and actual social and

---

[22]See http://new.paho.org/hq/dmdocuments/2011/dengue_cases_2010_May_20.pdf (accessed November 15, 2012).

political disruptions suggests that the following four factors are important: coping capacity; response capacity; surge capacity; and the likelihood of effective response. Governments of affected countries, as well as other response organizations, do not always use their response capacity fully and effectively. Some are more likely than others to have an effective response. This has implications both for response to particular events in the short term and for longer-term social and political stability.

Together these factors provide a foundation for understanding and assessing the likely short-term social and political effects of climate-related events. They also interact with other basic social and political conditions, including the levels and forms of political opposition, to affect the likelihood of recovery over the longer term.

General Approaches to Monitoring

One way to think about the monitoring needs for social and political stresses caused by climate-related events specifically is to imagine a nation-state affected by a major climate-related disaster situation in which social and political stresses would *not* become serious enough to undermine or topple the government, destabilize the regime, or threaten the stability of the state itself. A key to a government having this level of stability in the face of disaster is the extent to which there is both broad and deep legitimacy for the state, the regime, the government of the time, and the leadership group. A state with a high level of legitimacy would have, in the period before the event, no significant portion of the general population and no specific groups that reject with potential or actual violence (a) the existence or current territorial configuration of the country; (b) the form of governing the national territory, that is, the "regime"; (c) the right of the current government or administration to in fact govern that territory; or (d) the methods or processes by which the incumbent authorities occupy their positions. In the context of disruptive events, it is useful to think of legitimacy as providing a public reservoir of support (or at least acceptance) for a regime, a particular government, or a leadership group when a natural event becomes, through the combination of human system exposures and vulnerabilities, a major disaster. The Netherlands, which has obviously significant exposure to natural hazards and particularly to hazards associated with climate change, is an exemplar of a state and government unlikely to be destabilized by a disaster.

By contrast, a state with a political system with weak legitimacy—and therefore one that may be vulnerable in the case of a climate-driven natural disaster—would be one in which a significant portion of the general population or specific groups reject with potential or actual violence (a) the configuration of the national territory (e.g., in the case of secessionist, separatist, or independence movements), (b) the right of the current government or administration to rule (as evidenced by insurrections, insurgencies, or, in the case of opposition to authoritarian regimes, popular or democratic movements), or (c) the methods or processes by which the incumbent authorities came to be holding office (as evidenced by, on an increasing scale of violence, anti-government protests, demonstrations, or riots). The situation most likely to lead to problems would be a combination of all three rejections by multiple groups that agree on short-term goals (even if they differ on long-term goals), including at a minimum the ouster of the current authority figures. Such conditions would be conducive to the collapse or fall of the government or administration and possibly even to a regime change in the aftermath of the disaster. As discussed in Chapter 4, the case of Myanmar after 2008's Cyclone Nargis, where a type of slow democratization is currently occurring, is a possible example, depending upon the final outcome of the process.

Many countries at serious risk of disasters, including several of national security interest to the U.S. government, are in the substantial and complicated middle area between these two ideal types of high-legitimacy and low-legitimacy regimes. Turkey, for example, is closer to the first type—a country in which a major natural disaster (e.g., a great earthquake affecting Istanbul) would likely not pose a threat to the regime (although perhaps to the incumbent authorities and possibly the government of the time). Pakistan, with a major earthquake or flood, could be closer to the second type because, although the regime survived devastating floods in 2010, the aftermath of such an event could provide the occasion for an Islamist or military usurpation of power that would change not only the authorities but also the government and, quite likely, the entire regime.

Considering that in the 21st century the standard for legitimacy in most countries is some form or approximation of a democracy, it is worth considering and monitoring the characteristics of a democratic political system that might make it resistant to erosion or collapse in the wake of a major disaster. That set would include (1) sustained levels of high socioeconomic development without stark inequities; (2) regularized and accepted mechanisms and channels to organize and express grievances; (3) low levels of violence, particularly political violence; (4) a sustained record of low public corruption; (5) full civil liberties, a free media, and regular and fair elections with peaceful turnovers of post-election positions; and (6) effective governance capacities from the national to the local level, including for extreme event emergency response and recovery.[23]

---

[23] An example of a regime's stability under disaster can be seen in the situation that arose in Germany in August 2002 when devastating floods affected large areas and many cities and towns in the eastern part of the country. Interestingly, before the floods Chancellor Gerhard

A nation-state whose characteristics were a mirror image of these would likely have its stability threatened by a major disaster. Such a state would have: (1) low to middle levels of socioeconomic development, with stark inequities; (2) restricted, ineffective, or nonexistent mechanisms and channels to organize and express grievances; (3) high levels of violence, either currently or in the recent past; (4) sustained high levels of public corruption; (5) restricted civil liberties and media, fraudulent or non-competitive elections, and problematic turnovers of governing authority; and (6) ineffective or very limited governance capacities from the national to the local level, including for extreme event emergency response and recovery.[24]

Specific Monitoring Needs

These considerations suggest some specific monitoring needs for assessing the short-, medium-, and long-term effects of a major disaster or a sequence of disasters on social and political stability. In addition to the four factors introduced at the beginning of the section, monitoring should also be aimed at the following baseline and social and political conditions:

1. The *coping capacities* of the principal groups affected directly and indirectly by the event or sequence of events and the differences in coping capacity among the principally affected groups. Some existing indicators of social capital may serve as indicators of coping capacity (Aldrich, 2012).
2. The general *response capacity*, which can be monitored in part by observing the *response capacities* of responsible formal organizations. Response capacity is partly a matter of budgets and supplies (e.g., emergency shelter materials, water purification equipment, and food stocks). It also encompasses logistical capacities, the availability and training of general response personnel, and access to specialized personnel (e.g., first responders or medical personnel). Because capacity is meaningful only in relation to the costs of providing assistance—some of which, such as food prices, can be volatile— the costs of price-volatile supplies and equipment should also be monitored.

Schröder was behind in the polls for the September 2002 elections. His opposition to the oncoming U.S. invasion of Iraq combined with an effective disaster response and then a carefully financed recovery plan brought him back electorally, and he won reelection, albeit with a much reduced majority (from 21 to 9) in the Bundestag.

[24]These pre-conditions characterized Haiti when it was struck by a major earthquake in 2010, except that the United Nations and the community of nongovernmental organizations were providing most of the services normally associated with a state—and all Haitians were well aware of it.

3. *Expectations of response* on the part of the general public, key socioeconomic and cultural groups, and disaster-affected populations. This includes expectations of government emergency response capabilities and effectiveness as well as perceptions of government capabilities and effectiveness versus those of other domestic actors (e.g., the military or religious organizations in many countries), nongovernmental organizations in general, or outside donors. Expectations provide an important baseline indicator in advance of extreme events and can be measured by survey methods, perhaps as part of the more general efforts to survey public attitudes toward government discussed below. We are unaware, however, of any efforts to do such measurement systematically across countries for disaster-related issues. *Perceptions of response* after a disruptive event can be assessed in the same way and compared with the expectations to provide an indicator of the perceived adequacy of the response. The key dimensions of response perception include the effectiveness, transparency, and honesty of the responding organizations.
4. *Surge capacity*, one indicator of which is the availability of designated standby personnel and the funds to support them in the event of an emergency.
5. The *likelihood of effective response*. A baseline assessment could be provided by examining the track records of government and other response organizations in providing support to the affected populations or areas both generally and after past disruptive events, the history of national governments in allowing outside resources to flow to affected areas, and the record of national and local governments in delivering normal services to affected or potentially affected areas.
6. The direct and indirect social and economic *impacts* of the disaster (numbers killed, injured, and made homeless as well as the numbers of those whose livelihoods have been destroyed or jeopardized). The monitoring of these impacts, which is done in the aftermath of an event, could be done by class, race, ethnicity, gender, geography, political orientation, and so on. These factors are most important to monitor at the geographic level of the event, rather than at the national level. We recognize that some of this information may be difficult to collect, particularly in countries where some of these issues are likely to be sensitive and governments are either reluctant to ask the questions or else to make the information public. In such cases, it may be necessary to rely on indirect measures.
7. The impacts of actual *response* and *recovery*, which can be monitored through a number of attitudinal variables. These variables range from attitudes about specific responses to a disaster to attitudes about more general political and social conditions and capaci-

ties. These are normally measured by public opinion surveys asking for approval ratings, but they can also be assessed by examining the treatment of the government in free mass media. The attitudinal variables that can be monitored to assess the impacts of response and recovery can include

- *Expectations* of government disaster recovery capabilities and effectiveness by the general public, by key socioeconomic groups, and in the disaster-affected population.
- *Perceptions of the effectiveness* of government efforts in disaster recovery compared with efforts of other actors, institutions, or donors, assessed in the general public, key socioeconomic groups, and disaster-affected populations.
- *Mobilization of disaster-related grievances* by location, degree, type of grievance, population subgroup, and apparent purposes or goals. This monitoring should be particularly sensitive to concentrations of grievances by region, class, race, ethnicity, religious orientation, or other pre-existing societal cleavage.
- *Attitudes toward incumbent authorities* (i.e., the leaders or leadership group) from local to national levels. This can be assessed by approval ratings collected from general public samples, key socioeconomic groups, and disaster-affected populations, and should be sensitive to the above sets of pre-existing societal cleavage.
- *Attitudes toward the government or administration,* measured by approval ratings among the general public, key socioeconomic groups, and disaster-affected populations from local to national levels. This monitoring should also be sensitive to the above sets of pre-existing societal cleavage.
- *Attitudes about the underlying legitimacy or "rightness" of the nation, the state, the current government, and the incumbent authorities,* monitored in the general public, key socioeconomic groups, and disaster-affected populations, and sensitive to the above pre-existing societal cleavages. In surveys or focus groups, legitimacy can be assessed directly by asking whether the current form of government is appropriate for solving national problems. It can be assessed indirectly by examining such indicators as provision of basic health and security services.

8. *Social or political instability* over the longer term. Potential indicators include

- *Pre-existing levels of internal violence* and political instability in the country of interest.

- *Pre-existing levels of violence in neighboring countries* and changes in those levels after the event or sequence of events in the country of interest.
- *Attitudes toward or approval of* the nation, state, current government, and incumbent authorities within key institutions (e.g., the military, private-sector commercial and industrial organizations, and churches or religious organizations and movements). This is normally assessed with standard forms of political analysis, although there may also be a role for monitoring general or specialized media.
- *Mobilization of opposition* by parties, groups, or movements that may attempt to take advantage of the post-impact disaster period and the recovery period to advance their interests and agendas. This monitoring should be particularly sensitive to differences between largely non-violent "in-system" opposition parties, groups, or movements versus "out-system" and more violence-prone entities, especially if the latter are organized along the lines of major pre-existing societal cleavages. This is assessed by standard forms of political analysis.
- The *repressive capacities* of the state and the degree to which these are enhanced, remain static, or are degraded by the climate event or sequence of disasters.

Examples of Monitoring Resources for Coping, Response, and Recovery

Response capacity. The increasing international attention in recent years to reducing disaster risks rather than simply responding when disasters occur has led to a number of major communication and coordination initiatives that also offer sources of information about national, regional, and international capabilities. For example, the UN International Strategy for Disaster Reduction Secretariat (UNISDR), as part of its initiative to create national platforms for disaster risk reduction (81 countries at the time of this report) is developing a database of national response capacity.[25] Another UNISDR initiative is the Integrated Research on Disaster Risk's Forensic Investigations of Disasters,[26] launched in partnership with the International Social Science Council and the International Council for Science. Its goals are (1) to provide a baseline of the current state of the science in integrated research on disaster risk in order to measure the effectiveness of multiple programs, (2) to identify and support a long-term

---

[25] See http://www.unisdr.org/partners/countries (accessed November 15, 2012).
[26] See http://www.irdrinternational.org/about-irdr/scientific-committee/working-group/forensic-investigations/ (accessed November 15, 2012).

science agenda for the research community and funding agencies, and (3) to provide a scientific basis to support policy and practice. All of these various activities are yielding data on, or at least relevant to, national disaster response capabilities and offer relatively inexpensive monitoring possibilities for the U.S. intelligence community.

Perceptions and attitudes. In addition to traditional political analysis and research, both inside and outside the government, there are a number of resources for data on public perceptions and attitudes. In particular, public opinion survey projects such as Afrobarometer,[27] the Americas Barometer survey from the Latin American Public Opinion Project,[28] Asian Barometer,[29] and Arab Barometer[30] are accumulating evidence about attitudes toward democracy, government performance, and a range of social and political issues.

Regime types and governance. The longstanding Polity project, now part of the Political Instability Task Force effort, provides data that support quantitative and comparative analysis of regime authority characteristics and transitions. The types of governing authority range from *fully institutionalized autocracies* through *mixed, or incoherent, authority regimes* (termed "anocracies") to *fully institutionalized democracies.* The "Polity Score" captures this regime authority spectrum on a 21-point scale ranging from –10 (hereditary monarchy) to +10 (consolidated democracy) (Polity 4 Project, 2012).

Assessing the quality of governance has become a major focus for development assistance in recent years. Hundreds of datasets have emerged from donor agencies, governments, nongovernmental organizations, think tanks, and academia. (For a review of some of the "uses and abuses" of governance indicators, see Arndt and Oman, 2006.) Often used for research purposes because of their independence and freedom from member government influences, the annual datasets from Transparency International and Human Rights Watch are readily accessible and include various indicators for governance. The Worldwide Governance Indicators Project, produced by a team of scholars at the Brookings Institution with funding from the World Bank, is an example of a widely cited set of indicators from the world of intergovernmental organizations. It reports aggregate and individual indicators for six dimensions of governance: voice and accountability, political stability and the absence of violence, government effectiveness, regulatory quality, rule of law, and control of corruption. As explained on the World Bank's

---

[27]See http://www.afrobarometer.org (accessed November 15, 2012).
[28]See http://www.vanderbilt.edu/lapop (accessed November 15, 2012).
[29]See http://www.asianbarometer.org/newenglish/introduction/default.htm (accessed November 15, 2012).
[30]See http://www.arabbarometer.org/about.html (accessed November 15, 2012).

website, "These aggregate indicators combine the views of a large number of enterprise, citizen, and expert survey respondents in industrial and developing countries. The individual data sources underlying the aggregate indicators are drawn from a diverse variety of survey institutes, think tanks, non-governmental organizations, and international organizations" (World Bank, 2012). The data cover 213 economies for the period 1996–2010. A number of donor agencies also support national assessments. For example, as part of its Democracy, Human Rights, and Governance Program, USAID occasionally funds surveys that capture public experience with and perceptions of corruption.

Political instability and violence. A number of these resources were already noted in Chapter 5 as part of the discussion on assessing the evidence base for connections between climate change and security risk. The resources include the Uppsala Conflict Data Program,[31] the Political Instability Task Force (Box 5-1), and the Peace and Conflict Instability Ledger (Hewitt et al., 2012). A description of major datasets on international conflict and cooperation is maintained as an online appendix to the Compendium Project of the International Studies Association.[32]

Cross-cutting projects. In addition to the Political Instability Task Force, another U.S. government effort to examine a number of factors that affect coping, response, and recovery is the USAID Alert List. Since 2004 the USAID has produced an annual report that ranks countries according to (1) their current level of fragility and (2) a forecast of their risk for conflict or political instability and then combines the two rankings to produce an assessment of the most vulnerable countries. The seventh report was produced in 2011. The report is based on quantitative analysis of unclassified, open source information, and although the actual list of rankings is not made public, details about the methodology and some general findings are available. For example, in 2009, 23 of the 29 most vulnerable countries were in Africa (Moore, 2010:4)

The concept of "fragility" refers to the extent to which interactions between state and society produce outcomes that are (1) effective and (2) legitimate; the USAID measure assesses states on these two dimensions independently, and the two assessments for a state are often very different.[33] Of the 33 political, security, economic, and social indicators used to assess fragility, 16 are related to effectiveness and 17 to legitimacy. The forecasts of instability use a model developed by the Center for International De-

---

[31] See http://www.pcr.uu.se/research/UCDP/ (accessed November 15, 2012).
[32] See http://www.paulhensel.org/compendium.html (accessed June 23, 2102).
[33] Principal components analysis was used to calculate weightings based on the relative influence each indicator exerts on empirical relationships in the data, and those weights are then used to compute country scores.

velopment and Conflict Management at the University of Maryland. Its forecasts rely on a limited number of explanatory variables. In 2010 the project included an assessment of vulnerability to climate change based on a statistical model that analyzes global weather data and projects them forward. This approach permits physical climate hazards and human vulnerability to conflict to be disaggregated from the effects of climate change (Hewitt, 2012).

SUMMARY

The multitude of possible sources of data and indicators that could be applied to understanding climate-security connections is potentially overwhelming. This appendix attempts to provide an overview of major data sources and monitoring projects across the phenomena of interest of this project as well as a sense of the degree of consensus across research communities about key variables and indicators. It serves to illustrate the complexity and challenges associated with one of the report's findings: *Monitoring systems will require the integration of quantitative indicators of both environmental and social phenomena with traditional security and intelligence analytic methods.*

REFERENCES

Aldrich, D.P. 2012. The politics of natural disasters (pre-print). In *Oxford bibliographies in political science.* Available: http://works.bepress.com/cgi/viewcontent.cgi?article=1019&context=daniel_aldrich (accessed October, 4, 2012).
Arndt, C., and C. Oman. 2006. *Uses and abuses of governance indicators.* Paris, France: OECD, Centre for Development.
Barnett, J. 2006. Climate change, insecurity, and justice. Pp. 115–130 in *Fairness in adaptation to climate change,* W.N. Adger, J. Paavola, M.J. Mace, and S. Huq, Eds. Cambridge, MA: MIT Press.
Barnett, J., and W. N. Adger. 2007. Climate change, human security and violent conflict. *Political Geography* 26(6):639–655.
Brklacich, M., M. Chazan, and H.G. Bohle. 2010. Human security, vulnerability, and global environmental change. Pp. 35–52 in *Global environmental change and human security,* R. Matthew, J. Barnett, B. McDonald, and K. O'Brien, Eds. Cambridge, MA: MIT Press.
Brown, M.E., F. Tondel, T. Essam, J.A. Thorne, B.F. Mann, K. Leonard, B. Stabler, and G. Eilerts. 2012. Country and regional staple food price indices for improved identification of food insecurity. *Global Environmental Change* 22(3):784–794.
Burke, R.L., K.G. Vest, A.A. Eick, J.L. Sanchez, M.C. Johns, J.A. Pavlin, R.G. Jarman, J.L. Mothershead, M. Quintana, T. Palys, M.J. Cooper, J. Guan, D. Schnabel, J. Waitumbi, A. Wilma, C. Daniels, M.L. Brown, S. Tobias, M.R. Kasper, M. Williams, J.A. Tjaden, B. Oyofo, T. Styles, P.J. Blair, A. Hawksworth, J.M. Montgomery, H. Razuri, A. Laguna-Torres, R.J. Schoepp, D.A. Norwood, V.H. MacIntosh, T. Gibbons, G.C. Gray, D.L. Blazes, K.L. Russell, and AFHSC-GEIS Influenza Surveillance Writing Group. 2011. Department of Defense influenza and other respiratory disease surveillance during the 2009 pandemic. *BMC Public Health* 11(Suppl. 2):S6.

Cavallo, E., and I. Noy. 2010. *The economics of natural disasters*. IDB Working Paper Series No. IDB-WP-124. Washington, DC: Inter-American Development Bank.

Clinton, W.J. 1996. *Presidential decision directive NSTC-7*. Washington, DC: The White House.

Dawe, D. 2009. *The unimportance of "low" world grain stocks for recent world price increases*. ESA Working Paper No. 09-01. Food and Agriculture Organization. Available: ftp://ftp.fao.org/docrep/fao/011/aj989e/aj989e.pdf (accessed August 10, 2012).

Diaz-Quijano, F., and E.A. Waldman. 2012. Factors associated with dengue mortality in Latin America and the Caribbean, 1995–2009: An ecological study. *American Journal of Tropical Medicine and Hygiene* 86(2):328–334.

Famine Early Warning Systems Network. 2008. *FEWS NET strategy and vision for markets and trade*. Washington, DC: FEWS NET. (Updated January 2008). Available: http://www.fews.net/docs/special/FEWSNETMarketandTradeStrategy2005-2010.pdf (accessed December 17, 2012).

Fukuda, M.M., T.A. Klein, T. Kochel, T.M. Quandelacy, B.L. Smith, J. Villinski, D. Bethell, S. Tyner, Y. Se, C. Lon, D. Saunders, J. Johnson, E. Wagar, D. Walsh, M. Kasper, J.L. Sanchez, C.J. Witt, Q. Cheng, N. Waters, S.K. Shrestha, J.A. Pavlin, A.G. Lescano, P.C.F. Graf, J.H. Richardson, S. Durand, W.O. Rogers, D.L. Blazes, K.L. Russell, and AFHSC-GEIS Malaria and Vector Borne Infections Writing Group. 2011. Malaria and other vector-borne infection surveillance in the U.S. Department of Defense Armed Forces Health Surveillance Center-Global Emerging Infections Surveillance program: Review of 2009 accomplishments. *BMC Public Health* 11(Suppl. 2):S9.

Gleason, K.L., J.H. Lawrimore, D.H. Levison, T.R. Karl, and D.J. Karoly. 2008. A revised U.S. climate extremes index. *Journal of Climate* 21:2,124–2,137.

Gleick, P.H. (Ed.). 2011. *The world's water, Volume 7*. Washington, DC: Island Press.

Global Climate Observing System. 2003. *The second report on the adequacy of the global observing systems for climate in support of the UNFCCC*. April 2003 (GCOS-82, WMO/TD No. 1143). Available: http://www.wmo.int/pages/prog/gcos/Publications/gcos-82_2AR.pdf (accessed October 15, 2012).

Global Climate Observing System. 2010. *Implementation plan for the global observing system for climate in support of the UNFCCC* (2010 update). Geneva, Switzerland: World Meteorological Organization. Available: http://www.wmo.int/pages/prog/gcos/Publications/gcos-138.pdf (accessed August 6, 2012).

Hayden, M. 2012. The dengue vector mosquito *Aedes aegypti* at the margins: Sensitivity of a coupled natural and human system to climate change. Briefing to the Committee on Assessing the Impacts of Climate Change on Social and Political Stresses, January 12, Washington, DC.

Hewitt, J.J. 2012. *2011 alert lists: Methodological overview*. Briefing to the Committee on Assessing the Impacts of Climate Change on Social and Political Stresses, March 1, Washington, DC.

Hewitt, J.J., J. Wilkenfeld, and T.R. Gurr. 2012. *Peace and conflict 2012*. Boulder, CO: Paradigm.

Intergovernmental Panel on Climate Change. 2007. *Climate change 2007: Impacts, adaptation, and vulnerability*. Cambridge, UK: Cambridge University Press.

Intergovernmental Panel on Climate Change. 2012. *Managing the risks of extreme events and disasters to advance climate change adaptation*. Special report of working groups I and II of the Intergovernmental Panel on Climate Change. C.B. Field, V. Barros, T.F. Stocker, D. Qin, D.J. Dokken, K.L. Ebi, M.D. Mastrandrea, K.J. Mach, G.-K. Plattner, S.K. Allen, M. Tignor, and P.M. Midgley, Eds. Cambridge, UK: Cambridge University Press.

Johns, M.C., R.L. Burke, K.G. Vest, M. Fukuda, J.A. Pavlin, S.K. Shrestha, D. Schnabel, S. Tobias, J.A. Tjaden, J.M. Montgomery, D.J. Faix, M.R. Duffy, M.J. Cooper, J.L. Sanchez, D.L. Blazes, and AFHSC-GEIS Outbreak Response Writing Group. 2011. A growing global network's role in outbreak response: AFHSC-GEIS, 2008–2009. *BMC Public Health* 11(Suppl. 2):S3.

Koffi, J.K., P.A. Leighton, Y. Petcat, L. Trudel, L.R. Lindsay, F. Milford, and N.H. Ogden. 2012. Passive surveillance for *I. scapularis* ticks: Enhanced analysis for early detection of emerging Lyme disease risk. *Journal of Medical Entomology* 49:400–409.

Lagi, M., K.Z. Bertrand, and Y. Bar-Yam. 2011. *The food crises and political instability in North Africa and the Middle East.* Available: http://arxiv.org/pdf/1108.2455v1.pdf (accessed October 4, 2012).

Leichenko, R.M., and K.L. O'Brien. 2008. *Environmental change and globalization: Double exposures.* New York: Oxford University Press.

Lindgren, E., Y. Andersson, J.E. Suk, B. Sudre, and J.C. Semenza. 2012. Monitoring EU emerging infectious disease risk due to climate change. *Science* 336:418–419.

Money, N.N., R. Maves, P. Sebeny, M. Kasper, M.S. Riddle, and AFHSC-GEIS Enteric Surveillance Writing Group. 2011. Enteric disease surveillance under the AFHSC-GEIS: Current efforts, landscape analysis and vision forward. *BMC Public Health* 11(Suppl. 2):S7.

Moore, F. 2010. *Climate change in Africa.* Testimony before the Subcommittee on African Affairs and Global Health, Committee on Foreign Affairs, United States House of Representatives, April 15. Available: http://gopher.info.usaid.gov/press/speeches/2010/ty100415.html (accessed July 9, 2012).

National Research Council. 2001. *Under the weather: Climate, ecosystems, and infectious diseases.* Committee on Climate, Ecosystems, Infectious Diseases, and Human Health. Washington, DC: National Academy Press.

National Research Council. 2010a. *Advancing the science of climate change.* Panel on Advancing the Science of Climate Change. Washington, DC: The National Academies Press.

National Research Council. 2010b. *Monitoring climate change impacts: Metrics at the intersection of the human and earth systems.* Committee on Indicators for Understanding Global Climate Change. Washington, DC: The National Academies Press.

National Research Council. 2010c. *Adapting to the impacts of climate change.* Panel on Adapting to the Impacts of Climate Change. Washington, DC: The National Academies Press.

National Research Council. 2012. *Earth science and applications from space: A midterm assessment of NASA's implementation of the decadal survey.* Committee on the Assessment of NASA's Earth Science Program. Washington, DC: The National Academies Press.

O'Brien, K., and R. Leichenko. 2007. *Human security, vulnerability, and sustainable adaptation.* New York: United Nations Development Programme. Available: http://hdr.undp.org/en/reports/global/hdr2007-8/papers/O'Brien_Karen%20and%20Leichenko_Robin.pdf (accessed August 10, 2012).

Office of the Director of National Intelligence. 2012. *Global water security: Intelligence community assessment.* Washington, DC: Office of the Director of National Intelligence.

Ogden, N.H., C. Bouchard, K. Kurtenbach, G. Margos, L.R. Lindsay, L. Trudel, S. Nguon, and F. Milord. 2010. Active and passive surveillance and phylogenetic analysis of *Borrelia burgdorferi* elucidate the process of Lyme disease risk emergence in Canada. *Environmental Health Perspectives* 118:909–914.

Paavola, J. 2008. Livelihoods, vulnerability and adaptation to climate change in Morogoro, Tanzania. *Environmental Science and Policy* 11:642–654.

Palmer, W.C. 1965. *Meteorological drought.* Washington, DC: U.S. Department of Commerce.

Parry, M., N. Arnell, P. Berry, D. Dodman, S. Fankhauser, C. Hope, S. Kovats, R. Nicholls, D. Satterthwaite, R. Tiffin, and T. Wheeler. 2009. *Assessing the costs of adaptation to climate change: A review of the UNFCC and other recent estimates.* London, UK: International Institute for Environment and Development and Imperial College of London, Grantham Institute for Climate Change.

Polity 4 Project. 2012. *Polity 4 Project: Political regime characteristics and transitions, 1800–2010.* Available: http://www.systemicpeace.org/polity/polity4.htm (accessed July 28, 2012).

Russell, K.L., J. Rubenstein, R.L. Burke, K.G. Vest, M.C. Johns, J.L. Sanchez, W. Meyer, M. Fukuda, and D.L. Blazes. 2011. The Global Emerging Infection Surveillance and Response System (GEIS), a U.S. government tool for improved global biosurveillance: A review of 2009. *BMC Public Health* 11(Suppl. 2):S2.

Singh, R.P., D.P. Hodson, J. Huerta-Espino, Y. Jin, P. Njau, R. Wanyera, S.A. Herrera-Foessel, and R.W. Ward. 2008. Will stem rust destroy the world's wheat crop? *Advances in Agronomy* 98:271–310.

Tatem, A.J., S. Adamo, N. Bharti, C. Burgert, M. de Castro, A. Dorelien, G. Fink, C. Linard, J. Mendelsohn, L. Montana, M. Montgomery, A. Nelson, A.M. Noor, D. Pindolia, G. Yetman, and D. Balk. 2012. Mapping populations at risk: Improving spatial demographic data for infectious disease modeling and metric derivation. *Population Health Metrics* 10(1):8.

Trenberth, K.E., R. Anthes, A. Belward, O. Brown, E. Haberman, T.R. Karl, S. Running, B. Ryan, M. Tanner, and B. Wielicki. 2012. Challenges of a sustained climate observing system. Paper presented for WCRP Open Science Conference 2011, Denver, CO, October 24–28. (Revised February 15, 2012). Available: http://www.cgd.ucar.edu/cas/Trenberth/trenberth.papers/Trenberth%20paper%20OSC%20October%202011_v13.pdf (accessed August 5, 2012).

United Nations. 2012. *The Millenium Development Goals report.* New York: United Nations.

Witt, C.J., A.L. Richards, P.M. Masuoka, D.H. Foley, A.L. Buczak, L.A. Musila, J.H. Richardson, M.G. Colacicco-Mayhugh, L.M. Rueda, T.A. Klein, A. Anyamba, J. Small, J.A. Pavlin, M. Fukuda, J. Gaydos, K.L. Russell, and AFHSC-GEIS Predictive Surveillance Writing Group. 2011. The AFHSC-Division of GEIS Operations Predictive Surveillance Program: A multidisciplinary approach for the early detection and response to disease outbreaks. *BMC Public Health* 11(Suppl. 2):S10.

World Bank. 2012. *World governance indicators.* Available: http://info.worldbank.org/governance/wgi/index.asp (accessed July 28, 2012).

Wright, B.D. 2011. The economics of grain price volatility. *Applied Economic Perspectives and Policy* 33(1):32–58.